Pèlerin de Prusse on the Astrolabe

medieval & renaissance
texts & studies

Volume 127

Pèlerin de Prusse on the Astrolabe

Text and Translation of his
Practique de astralabe

by

Edgar Laird ■ Robert Fischer

Medieval & Renaissance texts & studies
Binghamton, New York
1995

A grant from Southwest Texas State University
has helped meet production costs of this book.

© Copyright 1995
Center for Medieval and Early Renaissance Studies
State University of New York at Binghamton

Library of Congress Cataloging-in-Publication Data

Pèlerin de Prusse, 14th cent.
 [Practique de astralabe. English]
 Pèlerin de Prusse on the astrolabe / text and translation of his Practique de astralabe / [edited] by Edgar Laird, Robert Fischer.
  p. cm.— (Medieval & Renaissance texts & studies ; v. 127)
 Includes bibliographical references.
 ISBN 0-86698-132-2
 1. Astrolabes—Early works to 1800. I. Laird, Edgar, 1937- . II. Fischer, Robert. III. Title: On the astrolabe. IV. Series.
QB85.P45 1994
522'.4—dc20   94-8986
                                                                             CIP

This book is made to last.
It is set in Baskerville, smythe-sewn
and printed on acid-free paper
to library specifications.

Printed in the United States of America

Cover design from a drawing in the Early English Text Society edition of Chaucer's *Astrolabe* (Oxford, 1872), captioned "W. W. Skeat fecit."

Table of Contents

Introduction	1
Pèlerin de Prusse	1
Astrology at the Court of Charles V	5
Scientific Writing in French	8
Subject of the *Practique*	10
Pèlerin's Source and Chaucer's	13
Pèlerin's *Practique de astralabe* and Chaucer's *Treatise on the Astrolabe*	17
Pèlerin's Language and Style	21
Technical Terminology	23
The Manuscript	25
Text and Translation	28
Practique de astralabe	31
Textual Notes	64
Explanatory Notes	65
Appendices	81
A. Excerpt from Messahalla's treatise on the astrolabe	83
B. Diagrams of parts of the astrolabe	88
C. Excerpts from Pèlerin's treatise on astrological elections	91
Bibliography	105
Index of Terms	113

Introduction

The *Practique de astralabe*, as it is entitled in its *explicit*, is a medieval French treatise on the ancient astronomical and astrological instrument called the astrolabe. Written by Pèlerin de Prusse in 1362 at the behest of Charles V (while he was still dauphin), the *Practique* is an important example of early technical and scientific writing in the vernacular and offers illustrative indications of the court's intellectual occupations and abilities. Since the treatise is heavily dependent on the same source as that on which Geoffrey Chaucer relied for his more famous work on the same subject, the *Treatise on the Astrolabe*, it is also interesting for the comparisons it invites with Chaucer's work. The present study offers an edition and translation of the *Practique*.

Pèlerin de Prusse

What little is known of Pèlerin de Prusse derives from his association with the royal court and from his writings. He appears in court records in the early 1360s as astrologer and "beloved clerk" to Charles, Duke of Normandy, later Charles V.[1] It is recorded also that when Charles was establishing the Hôtel Saint-Pol as his principal residence (1360-1362), he provided living arrangements there for "Master Pèlerin the astrologer and his man-servant."[2]
Contemporary and near-contemporary testimony states that Charles loved the sciences of the stars and those who professed them, and that he caused to be translated into French all the Latin treatises on the subject he could discover.[3] Pèlerin, at Charles's command, appears to

---

[1] "... notre amé clerc maistre Pelerin de Prusse." The record, dated Nov. 1, 1361, is quoted by Robert Delachenal, *Histoire de Charles V* (Paris: Librairie Alphonse Picard et Fils, 1909), 2:279.

[2] "Me Peregrin, astronomian et son valet," quoted in Delachenal, *Histoire*, 2:295. Lynn Thorndike, *A History of Magic and Experimental Science* (New York: Columbia Univ. Press, 1934), 3:586, lists the following forms of Pèlerin's name: Pelerinus de Prussia and Pelerin de Prusse, Pruce, or Pousse.

[3] Christine de Pizan, whose father Thomas de Pizan or Thomas of Bologna became an official astrologer to Charles V in 1368, writes that Charles was a "bon astrologien, et celle science amoit" and that he "singulierement aimoit philosophes

have collaborated in the enterprise of translation and thus produced, among other works, the *Practique*, which is a translation of parts of the *Compositio et operatio astrolabii*, John of Seville's twelfth-century Latin version of a lost Arabic treatise attributed to the eighth-century writer Messahalla (Māshā'allāh).[4]

The *Practique* and one other work, the *Livret de eleccions*, are both attributed to Pèlerin in the manuscripts, and both provide, in an autobiographical way, some further information about Pèlerin and his status. In the prologue to the *Practique*, Pèlerin says he writes at Charles's command for the benefit of those whose Latin does not extend to reading about astrolabes in that language, and in the *explicit* he says he completed the work on May 9, 1362. In the *Livret*, written the preceding year, he is a little more forthcoming but also a good deal more rhetorical, especially in his expressions of humility. In the prologue he calls himself one of the least of those who profess the sciences of the stars, unworthy to write on the profundities of so difficult a subject because of his youth, ignorance, and inexperience. He says he is nevertheless compelled to do so by Charles (of whom he is unworthy, the least of his subjects) who has commanded him to write briefly and clearly in French, a language he scarcely knows. From all this we may perhaps extract three facts: Pèlerin is young, he writes at Charles's command, and French is not his native language.[5]

Against his self-deprecation must be set his rather audacious willingness to state, in the body of the *Livret*, a general rule governing astrological elections that he claims is the key to all other rules and that he says he has seen in no other book on elections.[6] Against his stated

---

en le science d'astrologie" (*Le Livre de la paix*, ed. Charity Cannon Willard [The Hague: Persea Books, 1958], 142; also 207, note, quoting Christine's *Corps de pollicie*). In her biography of Charles, Christine writes a chapter on "comment le roy Charles estoit astrologien, et que est astrologie" (*Livre des fais et bonnes meurs du sage roy Charles V*, ed. S. Solente [Paris: H. Champion, 1940], 2:15–19). Simon de Phares, a fifteenth-century apologist for astrology, says of Charles V, "Cestui ayma tant la science de astrologie qu'il fist translater tous les livres qu'il peut finer et trouver de la science des estoilles...." (*Recueil des plus célèbres astrologues et quelques hommes doctes*, ed. Ernest Wickersheimer [Paris: H. Champion, 1929], 228).

[4] The attribution to Messahalla now appears doubtful (Paul Kunitzsch, "On the Authenticity of the Treatise on the Composition and Use of the Astrolabe Ascribed to Messahalla," *Archives internationales d'histoire des sciences* 31 [1981]: 42–62), but for convenience we continue to refer to the author as Messahalla.

[5] Oxford, St. John's College, MS. 164, fol. 33v. The passage is printed in Thorndike, *History* 3:587, n. 12.

[6] Oxford, St. John's College, MS. 164, fol. 110r.

reliance on "the wise men of ancient times"[7] must be set his account of "a new instrument" that he says is better even than an astrolabe for making elections.[8] Pèlerin was, and clearly regarded himself as, very accomplished in his art and well established with his royal patron when he completed the *Livret* on July 11, 1361 in the small "conserge-rie" of the Hôtel Saint-Pol.[9] His level of learning should warn us that, although the *Practique* can be regarded as a fairly close translation from Messahalla, Pèlerin was not utterly dependent on his source but could from time to time draw upon a body of knowledge he had accumulated elsewhere.

The *Practique* is preserved, so far as we can discover, only in Oxford, St. John's College, MS. 164, fols. 111-118. The *Livret* is preserved in the same volume, at fols. 33-110, and also in Rome, Biblioteca Vaticana Regina Sueviae 1337, fols. 45-88. The St. John's volume contains two other works, unattributed in the manuscript, that are sometimes said to be by Pèlerin.[10] One of them, however, the *Traité*

---

[7] Ibid., fol. 33v.

[8] Ibid., fol. 73v: "Et pour ce [i.e., to find the location of planets] faut il avoir au moins un astrolabe ou un autre instrument plus convenable, pour quoy un nouvel instrument est ordené par le quel legierement se puet faire ceste ouvrage." The instrument is probably an equatorium, which is by definition an instrument that permits one to determine the location of planets. See Emmanuel Poulle, *Les Instruments de la théorie des planètes selon Ptolémée* (Geneva: Librairie Droz, 1980), 1:12. It is not really a new instrument at the time but rare in the West. See "History of the Planetary Equatorium" in D. J. Price's edition of *The Equatorie of the Planetis* (Cambridge: Cambridge Univ. Press, 1955), 119-33. Pamela Robinson argues that Chaucer wrote the *Equatorie of the Planetis* in 1393 and in writing about an equatorium at that time "was very *avant garde*" (*Chaucer Review* 26 [1991]: 27). Simon de Phares says that when Charles V founded at Paris a college of astrology and medicine he provided it with both astrolabes and equatories (*Recueil*, 228). He also reports the remarkable construction of an equatorium in the year before Pèlerin wrote the *Livret*: Johannes de Santo Archangelo, he says, made an "equatoire des planetes en l'an 1360, qui est euvre de moult grant secours et utilité...." (*Recueil*, 227).

[9] Oxford, St. John's College, MS. fol. 110v: "... cest livret des ellecions ... je ai accompli par l'aide de Dieu a mon pouvoir l'an de grace 1361, le 11ᵉ jour de juillet, ascendant le 15 degré de Libre, le souleil a midi 4ᵉ etc. En la petite consergerie de l'ostel de mon seigneur de Normandie, de costé saint pol lez Paris."

[10] The treatise on the sphere at fols. 1-32 is attributed to Pèlerin by Claire Richter Sherman, *Medievalia et Humanistica*, n.s., 2 (1971): 87. The translation of Alchabitius at fols. 119-160 is attributed to Pèlerin by Moritz Steinschneider, *Europäischen Übersetzungen* (1904-1905, repr. Graz: Akademische Druck-u. Verlagsanstalt, 1956), 45-46, and *Hebräischen Übersetzungen* (1893, repr. Graz: Akademische Druck-u. Verlagsanstalt, 1956), 561-62; F. J. Carmody, *Arabic Astronomical and Astrological Sciences in Latin Translation* (Berkeley: Univ. of California Press, 1956), 148 (following Steinschneider); and J. D. North, *Chaucer's Universe* (Oxford: Oxford Univ. Press, 1988), 44, 48. George Sarton, *Introduction to the History of Science* (Baltimore: Carne-

de l'espere, is certainly by Nicole Oresme.[11] The other is a close translation of a treatise famous in the fourteenth century: John of Seville's Latin version of the *Introductory to Astrology* by the tenth-century Arabic writer Alchabitius (al-Qabīṣī). The content of Pèlerin's *Livret* strongly suggests that he knew Alchabitius's work, but then so did virtually everyone else in the fourteenth century who could lay claim to astrological learning.[12] Whether Pèlerin is responsible for the French translation must for now, in the absence of more definite evidence, remain a matter for speculation. But it seems to us that there is very little reason to think he was.

It is sometimes assumed that Pèlerin produced other works for Charles, and one suggestion is that these are among the items in Paris, Bibliothèque Nationale, ms. français 1083.[13] They are a book of "jugemens d'astrologie selon Aristote" and French versions of Zehel's *De judiciis astrorum* and Messahalla's *De receptione, De occulta*, and *Epistula de lunae eclipsis et planetis*. The suggestion is very tentative,

---

gie Institute, 1948), 3:1498, describes Pèlerin's treatise on elections and says it is derived or translated from Alchabitius; thus Sarton may have misled North, who seems to confuse the treatise on elections with the translation of Alchabitius. Henry O. Coxe, in his early description of the manuscript (*Catalogus* [Oxford, 1852], 2:52), queries whether an illegible word in the *explicit* to the translation from Alchabitius names Pèlerin. The word is indeed illegible, as if an attempt had been made to remove it, but enough is still visible to allow us to conclude that it is not "Pèlerin."

[11] See A. D. Menut, "A Provisional Bibliography of Oresme's Writings," *Mediaeval Studies* 28 (1966): 291, which, however, does not name the copy in the St. John's manuscript. The work is analyzed by G. W. Coopland, *Nicole Oresme and the Astrologers* (Cambridge, Mass.: Harvard Univ. Press, 1952), 13–20.

[12] Alchabitius's treatise, extant in more than a hundred copies in Latin, was a standard work in judicial astrology. It and a commentary on it by John of Saxony were given by Charles to the college of Maître Gervais at Paris. See Richard Lemay, "The Teaching of Astronomy in Medieval Universities, Principally at Paris in the Fourteenth Century," *Manuscripta* 20 (1976): 201–2. Alchabitius is quoted and named "Alkabucius" in Chaucer's *Astrolabe* I.8. A late fourteenth-century English translation of Alchabitius's treatise is in Cambridge, Trinity College, MS. 0.5.26.

[13] A. D. Menut, "Introduction" to Menut and Denomy, ed. and trans., *Nicole Oresme: Le Livre du ciel et du monde* (Madison: Univ. of Wisconsin Press, 1968), 5–6. In addition Menut mistakenly says that Delachenal, in "Note sur un manuscrit de la bibliothèque de Charles V," *Bibliothèque de l'Ecole des Chartes* 71 (1910): 33–38, attributes a similar collection to Pèlerin. The collection in question is Arsenal, MS. 2872, described in the *Catalogue des manuscrits, Bibliothèque de l'Arsenal* (Paris, 1885–99), 2:135–38. One treatise in that collection, "Le Livre de 9 anciens juges de astrologie," was translated at Charles's command from Latin into French in 1361 by Robert Godefroy. Delachenal's point is simply that Robert Godefroy and Nicole Oresme and Pèlerin de Prusse were all participants in a program of translating astronomy and astrology into French, a program that Charles initiated in 1361–62.

but the titles imply, at any rate, the character of other work which Pèlerin might have done for Charles. It is a character similar to that of works done for Charles by others: the French translations of Ptolemy's *Quadripartitum* and Abraham ibn Ezra's *Beginning of Wisdom* and the *Livre des neuf anciens juges de astrologie*.[14] And it is a character similar to that of the only two works we can say with assurance Pèlerin did write: the *Livret de eleccions* and the *Practique de astralabe*.

Astrology at the Court of Charles V

As Pèlerin was not the only writer producing French treatises and translations for Charles V, neither was he the only astrologer attached to Charles's court. From the fifteenth-century writer Simon de Phares we learn of several others.[15] One is Pierre de Valois, who made astrological predictions for the years 1358 and 1360, although just when he came into Charles's service is uncertain, for Simon says that he was "much valued by the English and afterward by Charles V for his knowledge of the stars."[16] Another is Gervais Chrestien, a stipendiary of Charles's and one so highly regarded that when Charles founded at Paris a college of astrology and medicine, sometime between 1362 and 1370, he named it for Gervais.[17] Yet another is André Sully, who cast nativities for Charles's sons Charles (1369), Louis (1369), and John (1370) and who passed from Charles's service to that of Bertrand du Guesclin on October 2, 1369, his transfer being apparently one of a number of gifts and honors Charles was bestowing on Bertrand at that time.[18]

From other sources we know about other court astrologers. Thomas de Pizan, also called Thomas of Bologna, who had been a professor of

---

[14] Simon de Phares, *Recueil*, 228, mentions Charles's commissioning the translation of the *Quadripartitum* (see Menut, *Mediaeval Studies* 28 (1966): 297-98) and the work of Abraham ibn Ezra, whose *Beginning of Wisdom*, in French translation, has been edited by Raphael Levy (Baltimore, 1939). The *Livre de neuf anciens juges* is the work by Godefroy identified in note 13, above.

[15] Simon's *Recueil* contains some factual errors, but Thorndike, in a chapter on "Astrology in the Later Fourteenth Century" (*History*, 3:285-601), makes judicious use of it.

[16] *Recueil*, 227: Pierre was "moult apprecié des Anglois et depuis du roy Charles le Quint pour la science des estoilles."

[17] *Recueil*, 223, 228, and see Lemay, *Manuscripta* 20 (1976): 200-201.

[18] *Recueil*, 232: "Cestui fut baillé audit messire Berthrand le 2ᵉ jour d'Octobre, quant l'espee de France lui fut baillée et qu'il fut fait conestable de France, pour ce qu'il fut saige et bien entendu astrologien."

astrology and medicine at the University of Bologna and was one of the most famous astrologers of his time, became astrologer to Charles in about 1364.[19] He held that position until Charles's death in 1380 and lingered on at the royal court until his own death in 1384 or 1385. Dominicus de Clavasio, who had been on the medical faculty at Paris from 1356–1357, became an astrologer at the royal court in 1368.[20] In addition to his astrological work Dominicus produced mathematical treatises, one of which, his *Practica geometria* (1346), emphasizes the art that Pèlerin treats in the closing sections of the *Practique*, namely altimetry, the art of measuring indirectly distances and heights that cannot be stepped off or measured with a cord.[21]

In 1357, according once more to Simon de Phares, Henry of Hesse delivered before the king and princes of France a sermon that was "very learned in the sciences of the stars."[22] Simon, himself an apologist for astrology, calls Henry a great "astrologien" (i.e., student of stars) but does not mention that while at Paris Henry wrote two treatises against the practice of judicial astrology, a *Tractatus contra astrologos coniunctionista* (1373) and a *Questio de cometa* (on the comet of 1378).[23] The example of Henry of Hesse reminds us that while there were astrologers in and around Charles's court, there were also "astrologiens" who can be described as anti-astrologers—the best and most famous of whom was Charles's counselor and former tutor, Nicole Oresme. Oresme's anti-astrological *Contra judiciarios astronomos et principes in talibus se occupantes* was probably written in about 1360,[24] just when Charles was sponsoring treatises and translations on judicial astrology, including Pèlerin's *Livret de eleccions*. The *Contra judiciarios astronomos*, though written in Latin, is addressed not to clerics and scholars but to "princes and magnates," and Oresme's *Livre de divinacions* (ca. 1361–1365), apparently the first work he wrote in French, is an even more elaborate warning against the use of astrology by "princes and lords responsible for public governance."[25]

---

[19] Sarton, *Introduction* 3:1480–81; C. C. Willard, *Christine de Pizan: Her Life and Works* (New York: Persea Books, 1984), 20–21.

[20] Thorndike, *History*, 3:587–88.

[21] *Dictionary of the Middle Ages* (New York: Charles Scribner's Sons, 1982), 8:211 (s.v. "Mathematics"); Thorndike, "Vatican Manuscripts in the History of Science and Medicine," *Isis* 13 (1929): 69–70.

[22] *Recueil*, 223.

[23] Claudia Kren, "Homocentric Astronomy in the Latin West," *Isis* 59 (1985): 270.

[24] Menut, *Mediaeval Studies* 28 (1966): 288.

[25] *Tractatus*: "principes et magnates"; *Livre*: "princes et seigneurs auxquels

Finally, Oresme's *Traité de l'espere* (ca. 1366–1377), which occurs in the same manuscript with Pèlerin's *Practique* and *Livret de eleccions*, limits itself strictly to just so much description of the physical structure of the cosmos as is "honest for a man to know, especially a prince of noble mind." But, says Oresme, a prince should go no further:

> If he wanted to delve further into speculative judgments [i.e., speculations of judicial astrology], this would be for him a vain curiosity, a thing he must not busy himself with. If he wanted to dabble in judicial matters about fortunes to come, it would be a thing very uncertain and very presumptuous on his part and dangerous with respect to God and the world. He would put himself in peril of losing his body and soul and his wealth and honor, as I more fully explained and proved in a French book that I wrote on this subject.[26]

The French book referred to is undoubtedly the *Livre de divinacions*.

Clearly the study of the stars was important at the court of Charles V, whether one favored or opposed its judicial aspect. Lee Patterson has presented evidence and arguments suggesting that the presence of astrology in chivalric courts of the time was in fact almost inevitable. Certain structural contradictions in astrological theory, he argues, are tied to similar contradictions in chivalric ideology, thus producing a "chivalric obsession with astrology."[27] He shows that the obsession was evidenced in the late Middle Ages in princely courts throughout Europe but notes that the court of Charles V "had a special commitment to astrology."[28] Lynn Thorndike says that the astrologers and

---

appartient le gouvernement publique," quoted and discussed by Coopland, *Nicole Oresme*, 21–22.

[26] *Traité de l'espere*, fol. 30v: "Et se il se vouloit profunder plus avant quant a la speculative des jugemens, ce seroit curiosité quant a lui et chose ou il ne doit point mettre son entente, et se il en vouloit affectueusement savoir & enquirer quant a la practique des jugemens des fortunes avenir, ce seroit chose nient certaine, inpertinent a lui & perilleuse quant a Dieu & au monde, & se mettroit en peril de perdre ame & corps & biens & honneur, sicomme j'ai plus a plain declairé et prouvé en .1. livret en françois que j'ai fait a ceste propos et sus ceste matiere."

[27] Lee Patterson, *Chaucer and the Subject of History* (Madison: Univ. of Wisconsin Press, 1991), 215–22. Without attempting to summarize Patterson's arguments, we may say that a chivalric lord who tries to dominate crude matter (land, vassals, women) yet lives almost Platonically aloof from it is drawn to a science that dwells in the celestial causes in order to dominate earthly affairs. But the stars express inexorable universal law, and, as Patterson says, "witnessing simultaneously to a desire for control and a feeling of helplessness, astrology articulates with unusual economy the contradictions at the heart of aristocratic culture of the late Middle Ages."

[28] Ibid., 216–17. In fairness it should be added that Charles's court shows

the anti-astrologers perhaps appealed to different sides of Charles's nature, and no doubt they did.[29] But as Thorndike himself shows in another place, the side we now easily label "astrological" was then still difficult to separate from star-study in general, for it explained seasons and tides as well as natal destinies and other superstitions, and its relevance to medicine was admitted even by anti-astrologers such as Oresme.[30] If we are to understand astrology in something like the way Charles did, we must reflect that, as Thorndike says, "during the long period of scientific development before Sir Isaac Newton promulgated the universal law of gravitation, there had been generally recognized and accepted another and different universal natural law, which his supplanted. And that universal law was astrological." The fourteenth-century debates about the stars are not so much debates between two disciplines—astrology and astronomy, say—as they are debates within the discipline of star-study, which is about to split along the lines of its internal contradictions.[31] They are debates about the "dynamics of universal nature" and the "psycho-physical forces which course through human and non-human spheres of existence."[32] A search for astrological explanations of things was one aspect of a more general search for explanation that could be termed "scientific."

Scientific Writing in French

Given the interest in science shown in courtly culture, it seems in order to say a few words about science and vernacular language. A survey of treatises written in the fourteenth century shows that Latin was still the dominant language of science in Europe, but it appears to have been declining in relation to the vernaculars.[33] In the first half

---

greater commitment to learning in general, not just astrological learning, than other courts of the day.

[29] *History*, 3:585.

[30] "The True Place of Astrology in the History of Science," *Isis* 46 (1955): 273–78.

[31] In the Arabic "sciences of the stars" ('ilm al-nujūm), judicial astrology, or "the science of decrees of stars" ('ilm' aḥkām al-nujūm), was by the thirteenth century separated from astronomy, or the "science of the spheres" ('ilm al-falak). See Lemay, *Manuscripta* 20 (1976): 197–98, and George Saliba, "Astronomy/Astrology, Islamic" in the *Dictionary of the Middle Ages* 1:616.

[32] T. McAlindon, "Cosmology, Contrariety and the Knight's Tale," *Medium Aevum* 55 (1986): 47–48. The phrases, though written about Chaucer's Knight's Tale, apply to astrology in general in Thorndike's sense of the word.

[33] The numbers in this paragraph are taken from Sarton, *Introduction*, 3:1285–86.

of the century, Latin was used by 223 scientific authors, and in the second half it was used by 192. In the same century, scientific writing in all European vernaculars was on the increase, and the most striking gain was in French, which was used by twenty-two authors in the first half of the century and fifty-one in the second half. In the second half, French had become the leading European vernacular for science and represented a third of all vernacular scientific writing in Europe.

Particularly in the study of the stars, moreover, the character of science in French changed during the course of the century. French treatises written in the first half tend to draw mainly on Patristic sources, whereas those written in the second half give better representation to the more progressive Greco-Arabic science that had earlier been mostly confined to Latin and the schools.[34] A fourteenth-century scholar's foundation in astronomy would normally include the reading in Latin of an elementary *sphaera* such as that by John of Sacrobosco, Messahalla's treatise on the astrolabe, and a *theorica planetarum* such as the one attributed to a certain Gerard (not Gerard of Cremona) or the one written by Campanus of Novara.[35] Much of this foundational learning had been made available in French by the end of the century. Oresme's *Traité de l'espere* was a vernacular equivalent of Sacrobosco's *De sphere* and is actually superior to it, and Pèlerin's *Practique* supplied much of the substance of Messahalla's work on the astrolabe. The *theoricae* are somewhat more demanding and tedious, and no real equivalent of them appeared in French before the sixteenth century,[36] but some of the matter and method of the *theori-*

---

[34] Francis J. Carmody, ed., *Leopold of Austria: "Li Compilacions de le science des estoilles"* (Berkeley: Univ. of California Press, 1947), 38.

[35] Lynn Thorndike, *University Records and Life in the Middle Ages* (New York, 1944), 279–82, and "The Study of Mathematics and Astronomy in the Thirteenth and Fourteenth Centuries as Illustrated by Three Manuscripts," *Scripta Mathematica* 23 (1957): 67–76; James A. Weisheipl, "Curriculum of the Faculty of Arts at Oxford in the Early Fourteenth Century," *Mediaeval Studies* 26 (1964): 172–73, and "Developments in the Arts Curriculum at Oxford in the Early Fourteenth Century," *Mediaeval Studies* 28 (1966): 151–75; Richard Lemay, "The Teaching of Astronomy in Medieval Universities, Principally at Paris in the Fourteenth Century," *Manuscripta* 20 (1976): 197–217, esp. 210–15. Also the modern editions of the works in question: Lynn Thorndike, *The "Sphere" of Sacrobosco and Its Commentators* (Chicago: Univ. of Chicago Press, 1949); R. T. Gunther, *Chaucer and Messahalla on the Astrolabe* (Oxford, 1929); F. J. Carmody, *Theorica planetarum Gerardi* (Berkeley, 1942); Francis S. Benjamin, Jr., and G. J. Toomer, *Campanus of Novara and Medieval Planetary Theory* (Madison: Univ. of Wisconsin Press, 1971).

[36] Oronce Fine's *Theorique des cielz* (Paris, 1528) is described on the title page as newly and clearly rendered in French ("nouvelement et tres clerement redigee en

cae were provided in Book II of the fourteenth-century French version of Leopold of Austria's *Compilacions de lé sciences des estoilles*.[37] As the use of French expanded in the fourteenth century, it took in new territory and was tending toward a complete embrace of science, especially the science of the stars.

The Subject of the *Practique*

The opening words of the *Practique* identify its subject very precisely: "La science du firmament et du mouvement des estoiles en la partie de practique." The "science du firmament" is the broad discipline of star-study that under the name of *astronomia*, or sometimes *astrologia*, belongs to the quadrivium of university study.[38] Although *astronomia* is a unified discipline, it is divisible into two principal kinds: "Due sunt species astronomiae," says John of Saxony, writing in Paris in 1331.[39] One kind, he continues, concerns lines and circles measuring the movement of stars (Pèlerin's "mouvement des estoiles"), and the other, concerning the heavens' influence on earthly things, is called "ars iudiciorum astronomiae" or "ars iudiciorum astrologie" (in Pèlerin's *Livret* called "partie des jugemens en astrologie").[40] The study of the movement of stars is further divisible into two parts: the theoretical and the practical (Pèlerin's "partie de practique"). The theoretical part is geometrical, qualitative description that employs intellectual proofs. The practical part, which Pèlerin has identified as his subject in this work, applies the conclusions of the

---

langaige françois"). A fourteenth-century *theorica* does exist in English: the "Newe Theorike of Planetis ... after the Almagest of Ptholome" in Cambridge, Trinity College MS. 0.5.26, fols. 125–81.

[37] Carmody, in his edition of the *Compilacions*, discusses its relation to *theoricae* on 41–44.

[38] Weisheipl, *Mediaeval Studies* 26 (1964):172–73; Lemay, *Manuscripta* 20 (1976):197–98. On the term and concept *astronomia*, see Jim Tester, *History of Western Astrology* (New York: Ballantine Books, 1987), 19, 102–3.

[39] The quotation comes from the commentary on Alchabitius that Charles probably gave to the college of astrology and medicine he founded at Paris between 1362 and 1370. We quote from the edition of Venice, 1491, fol. 29. The division of *astronomia* into two kinds is common. See Lemay, *Manuscripta* 20 (1976):198 and cf. the remarks in a commentary on Sacrobosco's *De sphera*: "Prima ergo species astronomie est que de motu celi et circulorum eius.... Secunda vero species est etiam de motu celi et circulorum eius non absolute sed per comparationem ad diversos effectus in hoc mundo inferiori...." (Thorndike, ed., *The "Sphere" of Sacrobosco and Its Commentators*, 413).

[40] Oxford, St. John's College, MS. 164, fol. 39r.

theoretical part. The practical is important because it is, as Campanus of Novara says, "the immediate end of the science of proof and the necessary antecedent of the science of judgment."[41] Pèlerin is thus working in between the realms of geometrical astronomy and judicial astrology, at the point where *astronomia* begins to acquire obvious social and political significance.

In the *Practique*, however, he is quite reticent about just what that significance might be. He calls his explanation of how to cast a horoscope (*Practique*, II.16) the most important and most often necessary of all his explanations, but he does not say why it should be so. Casting a horoscope is fundamental to all branches of judicial astrology, and the astrolabe is adaptable to each branch, but these are points Pèlerin makes in the *Livret*.[42] In the *Practique* he is concerned solely with technique, strictly confining the treatise to its proper subject matter—the methodology that logically and pedagogically precedes the science of judgments. This order of study is laid out for Pèlerin's audience in the great handbook for princes, Egidio Colonna's *De regimine principum*, of which Charles and his circle possessed numerous copies in both Latin and French.[43] Noblemen, says Egidio, must have the geometry to understand the movement of stars, and they learn that part of *astronomia* readily enough because they wish to proceed to the judgment of stars, by which they know when to undertake works and begin battles.[44] Once again the astrolabe, and Pèlerin's *Practique de astralabe*, may be seen to lie between geometrical astronomy and judicial astrology. In the court of Charles V, Nicole Oresme represents an anti-astrological point of view,[45] and the Pèlerin of the *Livret* repre-

---

[41] *Theorica Planetarum* 2.40-41, 138 in Benjamin and Toomer, eds., *Campanus of Novara*.

[42] See Appendix C, selection from *Livret*.

[43] Richard F. Green, *Poets and Princepleasers* (Toronto: Univ. of Toronto Press, 1980), 140.

[44] Egidio, teaching what noblemen should study by example of the past, locates astronomia within the framework of the seven liberal arts: "La ·VI· si est geometrie qui enseigne les mesures et les quantitez des choses. Ceste science aprennoient les gentiz hommes por cen que astronomie, qui enseigne la quantité et la distincté et le cours des estoiles, ne puent estre parfetement seüe sanz geometrie. La ·VII· science si est astronomie, la quele les gentiz hommes par aventure aprennoient ça en arrieres por cen que il estoient trop curieus de savoir le jugement des estoiles, quer il ne voloient onques commencier euvres ne batailles jusqu'a tant que il seüssent tout le cours et les dispositions des esteiles, et cen siet hon par astronomie, par quoi il l'aprennoient volentiers," *Livre du gouvernement des rois*, ed. S. P. Molenaer (1899, repr. New York: AMS Press, 1966), 200.

[45] Oresme's general opposition to astrology, in the *Tractatus contra judiciarios*

sents an astrological one, as is sufficiently demonstrated by the excerpt from that treatise in Appendix C, below. But the Pèlerin of the *Practique* is safely on the margins of the dispute.

The planispheric astrolabe itself, the instrument for which the *Practique* serves as a user's manual, was one of the most important scientific instruments of the Middle Ages. Its origin and history have been sketched many times by modern historians, although some of the details are still being worked out. It is clear that the mathematical theory of stereographic projection, on which the astrolabe is based, was known at least as early as Hipparcus's time (ca. 150 BC), and many historians accept Otto Neugebauer's arguments that the astrolabe was known to Ptolemy (ca. AD 150) and that a treatise on the subject by Theon of Alexandrindinus (ca. 375) is preserved in substance in the writings of Johannes Philoponus (ca. 530) and Severus Sebokht (ca. 660).[46] Others are less certain that the theory had been put into full use in the construction of the instrument by Ptolemy's time; while locating the origin of the astrolabe in Hellenistic culture, they stress the role of Islam in perfecting it.[47] It is also clear that the astrolabe passed from Islam to the Latin West, though the details of how it did so are likewise still under study. Most recent work seems to support the earlier conclusion of J. M. Millás Vallicrosa that knowledge of the instrument entered through Catalonia, where contact between Islam and the West was continuous, in the late tenth century.[48]

---

astronomos and the *Livre de divinacions*, is well known. His mathematico-astronomical objections to astrology occur in his *De proportionibus proportionum* and *Ad pauca respicientes*, ed. Edward Grant (Madison: Univ. of Wisconsin Press, 1966). See Grant's "Introduction," 61–65.

[46] Neugebauer, "The Early History of the Astrolabe," *Isis* 40 (1949): 240–56; and see Ernst Zinner, "Cl. Ptolemaeus und das Astrolab," *Isis* 41 (1950): 286–87; the articles in the *Dictionary of Scientific Biography* (New York: Charles Scribner's Sons, 1970–80) by G. J. Toomer on "Ptolemy" and "Theon of Alexandria"; J. D. North, "The Astrolabe," *Scientific American* 230 (1974): 104; and Arno Borst, *Astrolab und Klosterreform* (Heidelberg: Carl Winter Universitätsverlag, 1989), 13–20.

[47] Emmanuel Poulle, *Les Instruments astronomiques du moyen âge* (Paris: A. Brieux-E. Poulle, 1983), 15; Harold N. Saunders, *All the Astrolabes* (Oxford: Senecio Publishing, 1984), 2; David C. Lindberg, *The Beginnings of Western Science* (Chicago: Univ. of Chicago Press, 1992), 264–66. Articles in the *Dictionary of the Middle Ages* (New York: Charles Scribner's Sons, 1982) by Silvio Bedini ("Scientific Instruments"), Owen Gingerich ("Astrolabe"), and George Saliba ("Astrology/Astronomy, Islamic") suggest that the astrolabe, in a form close to that known in medieval Europe, developed in Hellenistic culture just prior to being passed on to Islam in the seventh century.

[48] J. M. Millás Vallicrosa, *Assaig d'Historia de les Idees Fisiques i Matèmatiques a la*

The parts and uses of the astrolabe, described by Pèlerin himself, have been described by modern scholars so often and so well that it is unnecessary to describe them again in full here. However, Pèlerin wrote for an audience that had access to astrolabes, and therefore his treatise presupposes that the reader knows what one looks like. It is, in the form to which Pèlerin refers, a small hand-held instrument of which the basic part is a circular metal disc. On the back of the disc are a sighting rule and graduated scales that allow an observer to make fairly precise observations of celestial bodies and also to take altitudes of terrestrial objects such as towers. On the front of the disc is a hollowed-out depression into which fits a smaller disc engraved with celestial coordinate lines, and over that fits a star-map called a "rete." The parts of an astrolabe are shown in diagrams in Appendix B, below.

Some help in understanding the structure and function of the front side of the astrolabe is provided by an early fourteenth-century treatise on the celestial globe (i.e., a globe that carries a map of the sky instead of a map of the earth).[49] Both the astrolabe and the globe, says the treatise, are constructed in the likeness of the heavens, as if from an original. But we should conceive the astrolabe as a sphere spread out flat ("in plano extensam ymaginari"); hence the astrolabe is a little more tedious to use, for what is seen directly on the globe must be imagined indirectly on the astrolabe. The treatise thus refers in simple language to the stereographic projection of a sphere on a plane surface, mentioned above as the principle on which the astrolabe is constructed.

Pèlerin's Source and Chaucer's

The treatise on the astrolabe that we, following medieval practice, are attributing to Messahalla was re-edited at the beginning of the thir-

---

Catalunya Medieval (Barcelona, 1931), and see Poulle, *Instruments astronomiques*, 15; Borst, *Astrolab*, 21–30; J. D. North, review of Borst, *Astrolab*, in *Speculum* 67 (1992): 636–38; and Lindberg, *Beginnings of Western Science*, 268. The difficulties in identifying Western and Islamic texts on which to base a history of transmission are briefly discussed by E. Ph. Goldschmidt, *Medieval Texts and Their First Appearance in Print* (New York: Biblo and Tannen, 1969), 119–21.

[49] The treatise, which begins, "Totius astrologie speculatione radix...," is discussed by Richard Lorch, "The *sphaera solida* and Related Instruments," *Centaurus* 24 (1980): 153–61. Our paraphrase is from Oxford, Bodleian, MS Selden Supra 78, fol. 71.

teenth century to form what Ron B. Thomson calls "a kind of *corpus* of astrolabe literature containing most of the methods of astrolabe construction and use, as well as practice problems, which was widely used, especially for teaching."[50] It is this *corpus* that is Pèlerin's source and Chaucer's. Manuscript copies of it, however, are extraordinarily numerous and no modern critical edition of it has been published, so that the exact relation of Pèlerin's work to its source is sometimes doubtful. R. T. Gunther's and W. W. Skeat's editions of Messahalla, though not fully critical, are quite adequate for making comparisons with the *Practique* in a general way. But where Pèlerin differs in detail from the text in these editions, it is always possible that he is following a manuscript not represented in them.[51] A manuscript or class of manuscripts may have existed—and may still exist—that is closer to Pèlerin's source than is the text of a modern edition.

That such might indeed be the case is suggested by the fact that at some points Pèlerin and Chaucer are closer to one another than either is to Messahalla as represented by Gunther and Skeat. In the *Practique* II.11, for example, the first paragraph is very similar in detail to the *Astrolabe*, II.28, lines 1–11, but not so similar in detail to their common source, Messahalla's II.29, as represented by Gunther and Skeat. Pèlerin's second paragraph is similar in content to Chaucer's lines 22–35, and that content is altogether lacking in the modern editions of Messahalla. It is possible in principle that Chaucer read the *Practique*,[52] but since there seems to be no compelling reason to think he did, it may be suspected that Pèlerin and Chaucer knew versions of Messahalla that were, in some respects at least, similar to each other and different from what is printed in modern editions.

Michael Masi suggested some years ago that Oxford's Bodleian MS Selden Supra 78 contains the very copy of Messahalla that Chaucer

---

[50] *Jordanus de Nemore and the Mathematics of Astrolabes* (Toronto: Pontifical Institute of Mediaeval Studies, 1978), 24.

[51] On the inadequacies of the editions, see North, *Universe*, 45–46.

[52] The possibility is considered by North, *Universe*, 44–45. Chaucer was held captive in France from late in 1359 until March 1360, and he returned to France, probably as the English King's emissary, in October 1360. North concludes that though Chaucer may have met Pèlerin "it is unlikely that he acquired much real astronomical knowledge at this time." Whether Chaucer later read the *Practique* is, North says, "a matter of pure speculation." Comparison of the *Astrolabe* with the *Practique* yields results (reported in our notes on the *Practique*) that might be read as evidence of Chaucer's having read Pèlerin's work. It seems to us, however, to be less than conclusive, for the reasons given below.

used, and the suggestion has gained some acceptance.[53] If it is indeed Chaucer's exemplar, and if Pèlerin's exemplar is similar to it, then it should be a useful representation of Pèlerin's source. Our reasoning here is very speculative but is perhaps worth pursuing briefly. Much of Masi's case for the Bodleian manuscript as Chaucer's source rests on two marginal notes in the manuscript that may have been incorporated into Chaucer's text. One of them may have led to Chaucer's specifying, in *Astrolabe* II.1, that it is the *dorsum*, or "bakhalf," of the instrument that is used for the operation in question ("I soughte in the bakhalf of myn Astrelabie"). The same specification is in the corresponding section of the *Practique* (II.1: "sus le dos de l'instrument"). The other marginal note that Masi adduces may have led Chaucer to add, in *Astrolabe* II.34, the warning that "commoun tretes of the Astrelabie ne maken non excepcioun whether the mone have latitude or noon." To the corresponding passage of the *Practique*, Pèlerin adds a similar warning—that the procedure just outlined omits consideration of latitude.[54] Did Pèlerin use the same manuscript as Chaucer, or one containing the same marginalia, or one in which the matter in the notes is incorporated in the text?

There is a third marginal note in the Bodleian manuscript that may be relevant to these questions, this one appearing to produce an omission instead of an addition. Both Pèlerin, in *Practique* II.9, and Chaucer in *Astrolabe* II.18, follow Messahalla's II.16 quite closely, except that both omit Messahalla's statement of how to find the longitude of a whole declination by using a thread over the zodiacal pole of the astrolabe. The omission could have been prompted by a note, such as the one in the margin of the Bodleian manuscript, to the effect that the method will not work because the celestial sphere is fixed not over the poles of the zodiac but over the poles of the world.[55]

On the whole, however, the similarity of Pèlerin's and Chaucer's additions and omissions seems to weaken the case for either author's dependence on a particular manuscript. It suggests that competent

---

[53] Masi, *Manuscripta* 19:36–47. The *Riverside Chaucer*, ed. Larry D. Benson (New York: Houghton Mifflin Company, 1987), says that Masi "makes a very good case for Chaucer's having read Messahalla and other astronomical works in a particular manuscript, Bodleian Selden Supra 78; he ... traces two additions (*Astr.* 2.1 and 2.34) to the marginal notes in the Selden MS" (1092).

[54] See *Practique* II.14, last paragraph, and notes thereto.

[55] "Iste modus non cum omnio declinationis quia positio spere non cum super polos zodiaci sed super polos mundi" (fol. 66r).

astrolabists such as Pèlerin, Chaucer, and the Bodleian scholiast are each capable of independent thinking about the instrument and its uses. The Bodleian manuscript is not an especially good one, and, as may be seen by comparing *Practique* I and *Astrolabe* I with the excerpt from the Bodleian text reproduced in our Appendix A, both Pèlerin and Chaucer do a somewhat better job of naming and describing the parts of an astrolabe than does the Bodleian Messahalla.

There are, moreover, points where Pèlerin and Chaucer agree with one another and differ from both the Bodleian manuscript and the modern editions. Where Messahalla in all these versions mistakenly has the "circle" of Capricorn, Pèlerin and Chaucer correctly write "beginning" and "head" of Capricorn.[56] Where Messahalla has "hours" without specifying whether he means "equal" or "unequal" hours, Pèlerin and Chaucer add the necessary qualification.[57] Pèlerin and Chaucer both warn against basing calculations on the altitude of a body taken near the meridian. The warning, almost certainly, is not based on any version of Messahalla, but a little experience with an astrolabe makes the point obvious.[58] Finally, there are in Pèlerin and Chaucer shared departures from Messahalla by which they appear to be adapting their treatises to certain kinds of instruments. The astrolabes Charles V is known to have possessed varied in size, and some of them were quite small.[59] Pèlerin would have to take that fact into consideration. Chaucer, in his prologue, says that his description of the astrolabe is intended to acquaint "lite Lowys" with his "oune instrument," evidently a small one that Chaucer has given him. Both writers therefore refer to small "portatif" astrolabes on which not all almucantars are represented but only every other one or even few-

---

[56] See *Practique* I.16, and n. 41.

[57] See *Practique* I.13, and n. 34.

[58] See *Practique* II.3, and n. 62. In a paper presented at the Modern Language Association convention in 1976, Michael Masi reported that, having examined over 180 manuscript copies of Messahalla, he had found none containing the warning against taking the altitude of a celestial body near the meridian.

[59] Astrolabes are listed in the *Inventaire du mobilier de Charles V* (Paris, 1879) as follows: une estrellabe cuivre (item 1990); ung grant astralabe de cuyvre (2072); ung astralabe qui est de cuivre, ront (2216); deux astralabes de laton, l'ung plus grant que l'autre (2270); ung grant astralabe d'or pesant troys marcs, troys onces, quinze estellins (2714); ung tres petit astrelabe d'argent blanc pesant quinze estellins (2817); ung astralabe de cuivre et ung estuy de cuir (3119); ung grant astralabe d'argent blanc pesant six mars, six onces, dix estellins d'argent (3121). The inventory was taken at the end of Charles's life and so may include instruments he did not own in the 1360s.

er.⁶⁰ Also, both writers omit any reference to the "twilight line" of an astrolabe, a feature that marks the time of morning and evening prayer on Islamic astrolabes.⁶¹ It was often but not always marked on Western astrolabes and may have been absent from the ones Pèlerin and Chaucer had in mind.

Our conclusion is that where Pèlerin deviates from modern printed texts of Messahalla, he may in some cases be following a manuscript version we have failed to identify, but he (and Chaucer too) may be clarifying, correcting, and adapting his material out of his own knowledge and experience as well.

Pèlerin's *Practique de astralabe* and Chaucer's *Treatise on the Astrolabe*

Geoffrey Chaucer's *Treatise on the Astrolabe* (ca. 1391) has long been singled out for special attention as the earliest treatise on that subject to be written in English.⁶² Pèlerin's *Practique de astralabe* (1362) is the earliest such treatise in French. Each treatise consists of a prologue and two parts, one naming the elements of the astrolabe and the other describing some of the operations that can be performed with it. Chaucer announces in his prologue plans to write three other parts, but he appears actually to have written only the two parts that he, like Pèlerin, could take more or less directly from Messahalla's *Compositio et operatio astrolabii*. Part I of each treatise has as its main source a short introduction to Part II of Messahalla's work. (This introduction

---

⁶⁰ See *Practique* I.11, and n. 30, and I.12, and n. 32. The great majority of astrolabes range between 7 and 30 cm. in diameter. For classification and designation of astrolabes according to size and the number of almucantars represented (ranging from "perfect," with 90 almucantars, downward) see William H. Morley, *Planispheric Astrolabe* (London, 1856), 8.

⁶¹ Morley, *Planispheric Astrolabe*, 11.

⁶² For scholarship on Chaucer's *Astrolabe*, see Russell A. Peck, *Chaucer's* Romaunt of the Rose *and* Boece, Treatise on the Astrolabe, Equatorie of the Planetis, *Lost Works, and Chaucerian Apocrypha: An Annotated Bibliography, 1900 to 1985* (Toronto: Univ. of Toronto Press, 1988), 180–228. Additional recent studies are Jose-Luis Martinez-Duenas-Espejo, "La prosa cientifica de Geoffrey Chaucer: Estudio textual y gramatical de *A Treatise on the Astrolabe*," in *Estuidos literarios ingleses: Edad Media*, ed. J.-F. Galvan-Reula (Madrid: Catedra, 1985), 121-37; Marijane Osborne, "The Squire's 'Steed of Brass' as Astrolabe: Some Implications of the *Canterbury Tales*," in *Hermenutics and Medieval Culture*, ed. Patrick J. Gallacher (Albany: SUNY Press, 1989), 121-31; George Ovitt, Jr., "History, Technical Style and Chaucer's *Treatise on the Astrolabe*," in *Creativity and the Imagination: Case Studies from the Classical Age to the Twentieth Century*, ed. Mark Amsler (Newark: Univ of Delaware Press, 1987), 24-58.

is reproduced below as Appendix A.) The following table shows how the subsections of *Practique* I and *Astrolabe* I correspond with each other and with the subsections of the passage in Messahalla from which they derive. Key terms from each subsection are included.

Messahalla II, Introduction	Pèlerin I	Chaucer I
a. armilla suspensoria	3. annel, armille	1. ryng
b. ansa	2. une hautesse petite	2. toret
c. mater	4. mere	3. moder
margolabrum (elsewhere "limbus")	5. lymbe	16. bordure
d. circulus cancri	6. [cercle] du Cancre equinoctial	17. cercle of Cancre cercle equinoxiall
circulus equinoccalis		
circulus capricorni	cercle de Capricorne	cercle of Capricorne
e. almucantharat orizon	8. almucantharat orizon oblique	18. almycanteras orizonte
—	9. pote main, orientele l'autre moitié, occidentele	6. left syde, west ryght syde, est
f. cenit capitum	10. cenith	18. cenyth
—	11. la grandeur de l'instrument	18. the quantite of the Astrelabie
g. azimuth	12. azimuth	19. azimutz
h. hore	13. heures inequales	20. houres of planetes
—	14. pluseurs tables	—
i. crepusculorum linee	—	—
j. linea meridiei	7. lingne de midy	4. north lyne, lyne meridional
media noctis	lingne de minuit	south lyne, lyne of midnyght
—	lingne de vrai orient et occident	5. lyne orientale, lyne occidentale
k. alhancabuth, aranea via solis	15. reys, rethe ecliptique	21. rete ecliptik lyne
l. almuri, ostensor	16. almuri, monstreur	21. almury, denticle of Capricorne, calculer
m. foramen	17. trou	14. hole
axis	—	pyn, extre
equus	cheval	hors
cuneus	cheville	wegge

Messahalla II, Introduction	Pèlerin I	Chaucer I
n. in alterea parte matris	18. en l'autre partie	—
— (elsewhere "dorsum")	dos	4. bakhalf
2 circuli	2 manieres de cercles	—
numerum dierum, 365	le jour de l'an —	9. cercle of the daies, 365
nomina mensium	12 moys	10. the names of the monthes
graduum signorum	signes et leur degré	8. degrees of signes
nomina signorum	—	8. names of the 12 signes
o. quarta capiende altitudinis	—	—
p. quadrans in 12 puncta divisa	19. demy quarré partis en 12	12. 2 squyres 12 pointes
q. regula tabule perforata —	20. reigle tablet un trou ou 2	13. reule plate perced certein holes

The chief difference between Pèlerin's and Chaucer's handling of their common source is that Chaucer takes slightly greater advantage of the astrolabe's usefulness as a device for teaching astronomy. This is especially evident in *Astrolabe* I. 17 and 21, where he expands upon Messahalla's description of the rete of the astrolabe by adding material from John of Sacrobosco on the celestial zodiac, which the rete represents. Both Pèlerin and Chaucer omit Messahalla's item i, on the twilight lines (linee crepusculorum). Both writers add an explanation of which side of the astrolabe represents east and which west, although their usages differ (Pèlerin I.9; Chaucer I.6). And each adds a reference to the incompleteness of markings on smaller astrolabes (Pèlerin I.11; Chaucer I.18). In general, however, they both follow Messahalla quite closely in Part I of their treatises.

Neither Pèlerin nor Chaucer attempts to reproduce all of the subsections in Messahalla's Part II. Chaucer's *Astrolabe* II contains forty subsections (or perhaps as many as forty-six, depending on whether one accepts as Chaucer's any or all of the "supplementary propositions," namely II.41–46, at the end of the treatise). Pèlerin's *Practique* II contains only twenty. The following table, keyed this time to Pèlerin, shows the correspondences among the three works.

Pèlerin II	Messahalla II, Introduction	Chaucer II
1. signe et degré ... soleil dos	1. gradu solis —	1. degre ... sonne bakside
2. hauteur du soleil ... estoiles	2. altitudine solis et stellarum	2. altitude of the sonne or of othre celestial bodies
3. heure signe et degré ascendent	3. hore inequalis signi ascendentis	3. tyme ascendent, or ellis horoscopum
4. crepuscule	4. crepusculo	6. crepuscules
5. quantité du jour et de la nuit	5. arcus diurni et nocturni	7. arch of the day
6. combien des heures equales sont passees	8. De numero horarum dei equalium preteritarum	11. hou many houres of the clokke ben passed
7. quantité d'une heure inequale par jour	6. De quantitate horarum diei inequalium	10. quantite of houres inequales by day
8. hauteur du soleil a chascun midi	8. De altitudine solis in meridie	13. altitude meridian
9. quel degré du zodiaque ... avec chascune estoile fixe	16. gradus stelle cum quo celum	18. degrees of longitude of fixe sterres
10. quel endroit	17. cenith	33. cenyth
11. signe se lieve	28. ascensionis signorum	27. ascensions of signes
12. estoiles fixes qui sont mises dedenz le rethe	30. stellarum ... positarum in astrolabio	—
13. vray midi et vrai orient	19. De quatuor plagis mundi	29. the 4 quarters of the world
14. lieu des estoiles fixes qui ne sont mie situés en astralabe	31. stellarum incognitarum non positarum in astrolabio	—
15. se les planetes sont septentrioneles ou meridioneles	35. De latitudine planetarum a via solis	30. latitude of planetes fro the wey of the sonne

Pèlerin II	Messahalla II, Introduction	Chaucer II
s'il sont retrogrades ou directes	35. De directione et retrogradacione planetarum	35. yf that eny planete be direct or retrograd
16. les 12 maisons	37. 12 domorum	36. houses
17. hauteur du soleil par les heures	—	—
18. Premiere partie ... aucune chose haute a quoy nous pourrons approuchier	45. altitudinis rei accessibilis	41. altitude of the tour
Seconde partie chose ... assise que nous ne pourrons approuchier	46. altitudine rei inacessibilis	42. that thou may nat come to the bas of the tour
19. mesurer ... longue de plainne terre	47. mensuracione plani	—
20. hauteur des choses par leur ombres	—	—

Of the subsections in Chaucer that have no equivalent in Pèlerin, nine do not derive from Messahalla (Chaucer's II.4, 12, 22, 23, 26, 32, 38, 39, 40), and at least one other (II.8) is missing from an otherwise good manuscript of Messahalla. Chaucer has six subsections on geographical topics (II.21-25, 39) that Pèlerin does not treat. Chaucer's II.4 and II.12 have no matching discussions in Pèlerin; each is a "special declaracioun" on an astrological topic.

Pèlerin's Language and Style

The French language at the time Pèlerin de Prusse wrote the *Practique de astralabe* is characterized by rapid changes and widespread variability. The texts of leading authors of the time, such as Deschamps and Froissart, show inconsistencies in language usage even within a single document.[63] Pèlerin's *Practique* provides abundant examples of the same kinds of inconsistencies. For example, the verb *prendre* is realized in its various forms as *prendre*, *prandre*, and *prenre*; also,

---

[63] Ferdinand Brunot, *Histoire de la langue française* (Paris: Librairie Armand Colin, 1966), 1:421.

prenons alternates with *prendon*, and *prendrons* with *prenrons*. The atonic vowel of *pouvoir* varies between *ou* and *o*, and the intervocalic consonants are written as *rr* or *vr*, leading to forms such as *pouvons*, *povons*, *povoir*, *pourrons*, *porrons*, *pouvrons*, and *pouvra*. The word *trou* is spelled *trou*, *tro*, and *troes*; the word *prochain* as *prochainne* and *prouchainne*; and the word *moitié* as *moitié*, *muté*, and *mutet*. The plural masculine forms of adjectives ending in *-al* are spelled variously as *generalz*, *egalz*, *principaus*, and *principaulz*. Words such as *elle*, *telle*, and *appelé* occur with both single and double consonants: *elle*, *ele*, *telle*, *tele*, *appelé*, *appellé*. The word *vaut* is spelled consistently *vault*, but *haut* is found both with and without the consonant *l*: *haut* and *hault*. The words *rond* and *voir* occur with and without *e*: *reont*, *ront*, *reondesse*; and *voir* and *veoir*. The letters *y* and *i* alternate freely in *pays* and *vrai*: *pays* and *pais*; *vrai*, *vray*, and *vraye*. *Fois* is spelled *fois* and *foiz*. Word usage is generally consistent, but the words *lonc* and *longueur* both refer to length and *hauteur* and *hautesse*, to height. Some variability in gender is noticeable in the text. The word *espace* is usually feminine but occurs twice as a masculine noun, while the noun *planete* is usually masculine but occurs once as a feminine noun.

As the Francien dialect of the Ile de France enjoyed increasing prestige and exerted greater influence across the country, the language of vernacular courtly documents, legal and political, continued to be subject to the ever-present influence of Latin. Legal clerks in the mid-fourteenth century sometimes mixed Latin and French within the same document and frequently used Latin calques.[64] Although Pèlerin was attached to the court and accustomed to using Latin, he avoids the direct use of Latin expressions for the most part in his *Practique*, though he does describe large unwieldy astrolabes as "trop pesans et non portatis" and describes early evening as "le vespertin crepuscle." Also, many technical terms associated with astrolabes come from a lengthy Arabic-Latin tradition and are adopted for use in the *Practique*.

As noted earlier, Pèlerin admits to an imperfect command of French; and Delachenal, a scholar thoroughly acquainted with the prose of Pèlerin's time, concurs that Pèlerin writes with the clumsiness of a man insufficiently familiar with the French language.[65] The clum-

---

[64] Brunot, *Histoire* 1:529.

[65] Delachenal, *Histoire* 2:368 Over one of Pèlerin's especially fulsome and elaborately allusive references to Charles, Delachenal exclaims, "Grandes et belles idées, quoique pauvrement exprimées!" (Ibid. 2:369).

siness to which Delachenal refers, however, is in the stilted and ornate prologue to the *Livret*, where Pèlerin may have been attempting to "latinize" his style under the influence of the notaries and royal secretaries with whom he associated.[66] In the *Practique* his stated objective is to write simple, concise French,[67] and in that objective he has better success. If the success is incomplete, it is so partly because he had no stable tradition of French technical prose in which to work.[68] On the contrary he was helping to invent one.[69]

Technical Terminology

The technical vocabulary associated with the astrolabe reflects the Islamic heritage of Latin Europe's knowledge of the instrument.[70] The earliest complete Latin treatises on the astrolabe, those of Llobet of Barcelona and his circle (late tenth century), appear to come directly from Arabic sources, and they retain from them the names of various parts of the instrument.[71] Raymond of Marseilles's treatise on the astrolabe (1141) is perhaps the earliest in the Christian West that is not a translation or close adaptation of an Arabic work, yet it too employs terminology from the Arabic.[72] John of Seville's Latin version of Messahalla, which is not only Pèlerin's source but also, on the evidence of the number of surviving manuscripts,[73] the most

---

[66] Willard, in her edition of Christine de Pizan's *Livre de la paix*, 50–51, describes Christine as making the same attempt decades after Pèlerin wrote. Willard's description of the latinizing features of Christine's style applies almost as well to those of Pèlerin's.

[67] *Practique*, prologue and *explicit*.

[68] See Carmody, ed., *Leopold of Austria*, 37–39.

[69] The astrolabe was fundamental to the professional astrologer's art, and Pèlerin, as a professional, certainly knew how to operate one. He must have turned to Messahalla not as a source of information about astrolabes but as a model of how to write about them.

[70] The Arabic terms for the parts of an astrolabe, with transliterations and Latin equivalents, are given in Morley, *Planispheric Astrolabe*, 8–21; see also Willy Hartner, "The Principle and Use of the Astrolabe" (1939), reprinted in Hartner's *Oriens-Occidens* (Hildesheim: Georg Olms Verslagbuchhandlung, 1968), 287–311.

[71] Harriet P. Lattin, "Lupitus Barchinonensis," *Speculum* 7 (1932): 61–62; Thomson, *Jordanus de Nemore*, 23. Much of this literature is printed in Millas Vallicrosa, *Assaig*, 271–335.

[72] Emmanuel Poulle, "Le Traité d'astrolabe de Raymond de Marseille," *Studi Medievali*, 3d ser., 5 (1964): 867–68. Poulle's edition of Raymond's treatise is printed at 874–904.

[73] See Carmody, *Arabic ... Sciences in Latin Translation*, 23–25. Unless otherwise

widely used Western work on the astrolabe, also employs an Arabic-derived vocabulary. The nomenclature of the astrolabe, including its Arabic component, was established early and persisted virtually unchanged throughout the active history of the astrolabe. When Pèlerin uses the words *alidade, almucantharat, almuri, cenith,* and *nadair*, he is simply adopting standard Arabic-Latin terminology that also appears in the Llobet treatises, in Raymond of Marseilles, and in the Latin Messahalla.[74] Pèlerin also appropriates standard Latin-based technical terms such as *armille* (*armilla*), *lymb* (*limbus*), *rethe* (*rete*), and *table* (*tabula*, in the technical sense of "a plate or disc of an astrolabe or other instrument"). He also translates, as with *mere* for *mater* (which in turn translates the Arabic *umm*), *cheval* for *equus*, and *monstreur* for *ostensor*. In all cases he gives the practical meaning of the term in the only way possible—by describing the part or function of the astrolabe to which the term refers.

For terminology belonging to star-study in general, Pèlerin likewise draws mainly upon the standard Latin vocabulary of the discipline, as with the words *angle, ascendent, declinacion, ecliptique, equinoctial, latitude, oblique orizon*. All of these were already in use in French[75] but were new or recondite enough so that Nicole Oresme included them among the "moz estranges" in the glossorial index to his *Traité de l'espere*, a work contemporary with Pèlerin's and apparently addressed to the same audience.[76] Chaucer too uses each of these expressions, in Anglicized form. But how much further it would be possible to go in the way of vernacularizing astronomy is suggested by a late-fourteenth-century English manuscript which gives "corners" for *angles*, "brede" for *latitude*, "evennyght" for *equinoctial*, and "croked" for *oblique*.[77] Pèlerin's

---

stated, we refer to the treatise in the edition of R. T. Gunther in *Chaucer and Messahalla*.

[74] A short fragment of an early European astrolabe text, copied at Reichenau in the late tenth or early eleventh century, (Borst, ed., *Astrolab*, 121–27) uses the Arabic-derived terms *alhidada, almukantarat, nadyr,* and *almeri*.

[75] All but *ecliptique* and *equinoctial* occur in the part of *Li Compilacions de le science des estoilles* (written before 1324) edited by Carmody.

[76] Coopland, *Nicole Oresme*, 184–85. In Oxford, St. John's College, MS. 164, Oresme's glossorial index immediately precedes Pèlerin's *Livret de eleccions*. In the same manuscript, incidentally, several sheets are left blank at the end of Part I of the *Livret*, with an invitation to the reader to make his own glossary: "Et je ai laissié ces 3 fueillez voiz afin se il plaist de exposer aucune chose que elle soit ci escriptes" (fol. 69v).

[77] Cambridge, Trinity College, MS. O.5.26, fols. 1v, 1v, 8r, and 32r, respectively. The first three are in an English translation of Alchabitius's *Introductory to Astrology*,

language, by contrast, is in the main stream of development in technical vernacular usage.

Another category of technical usage is the mathematical, and Pèlerin's practice there is somewhat different. One of the chief points of the astrolabe is that it allows one to reach solutions to complex problems in spherical geometry with little recourse to mathematics. Consequently, Pèlerin uses only a few mathematical terms and concepts, namely those dealing with counting, adding, and subtracting: *compter* (count), *adjouster* (add), *oster* (subtract), *demourant* (remainder). He avoids the term *arc* or *archon* or *arch* (Messahalla's "arcus"), although it was being used by others in both French and English at the time. And where Messahalla instructs one to divide ("diuide") an arc by fifteen, Pèlerin suggests instead that one move a pointer across the border of the astrolabe and count the number of times the pointer crosses fifteen degrees. When forced to divide, he writes *fois dedenz* (as in "quantes fois les pointes prises sont dedenz 12") or *combien de fois en* (as in "le nombre des pointes et combien de fois elles sont en 12").[78] Chaucer, for his part, employs a mathematical vocabulary that includes not only *adden* and *diminucioun* (subtraction) but also *dividen*, *divisioun*, *fraccioun*, *proporcioun*, *multiplicacioun*, and *arch*.[79] In the *Astrolabe* he addresses an audience that, as he says in the prologue, gives evidence of "abilite to lerne sciences touching nombres and porporciouns." Pèlerin, apparently, is not addressing such an audience. As he certainly understood the simple arithmetic he found in his source, his further simplification of it must be seen as a deliberate attempt to avoid technicality.

The Manuscript

The Parisian manuscript in which the *Practique* is preserved—Oxford, St. John's College, MS. 164—was given to St. John's in 1633 by William Paddy.[80] When and how it came to England is uncertain, but

---

and the last is in an English translation of William of England's *De urina non visa*.

[78] Both examples come from *Practique* II.18.

[79] These words were "only just coming into written use when Chaucer employed them" (Ralph W. V. Elliot, *Chaucer's English* [London: André Deutsch, 1974], 317–18). The French mathematical vocabulary, certainly by the fifteenth century, was much richer than Pèlerin's usage would suggest. See Paul Benoît, "Recherches sur le vocabulaire des opérations élémentaires dans les arithmétiques en langue française de la fin du moyen âge," *Documents pour l'histoire du vocabulaire scientifique* 7 (1985): 77–95.

[80] The manuscript is described by Henry O. Coxe, *Catalogus codicum MSS. qui in*

it is known to have been in France, in the royal study at Vincennes, in 1418.[81] The currently suggested date of it, "around 1377," is probably accurate, although some problems with dating are noted below. It measures 200 x 142 millimeters and contains 161 numbered vellum folios of clear writing. The contents are as follows:

1. Four preliminary, unnumbered leaves containing crude, mostly illegible horoscopes
2. Nicole Oresme, *Traité de l'espere* (fol. 1)
3. Pèlerin de Prusse, *Livret de eleccions* (fol. 33)
4. Pèlerin de Prusse, *Practique de astralabe* (fol. 111)
5. Alchabitius, *Introductoire* (fol. 119)
6. A table of contents for the volume, in a later hand (fol. 153)
7. Notes on astrological signs, in the first hand (fol. 155)
8. Illuminated horoscopes of Charles V and his children, Charles the dauphin (later Charles VI), Marie, Louis, and Isabelle (fol. 158)

There are four miniatures in the manuscript. At the beginning of Oresme's *Traité* (fol. 1r) is a crowned figure (Charles V) seated alone at a small octagonal desk.[82] Behind him is a book-press suspended on the wall. With his left hand he holds a book open on his lap. On the desk are an armilliary sphere (appropriate to the subject of Oresme's *Traité*) and more books, one of which he holds open with his right hand.

Also near the beginning of Oresme's *Traité* is a miniature of a standing, bare-headed figure (Oresme) holding an armilliary sphere and pointing to it as if giving instruction.[83]

---

collegiis aulisque Oxonienibus hodie adservantur, vol. 2 part 6 (Oxford, 1852), 51–52; Léopold Delisle, *Recherches sur la Librairie de Charles V* (Paris, 1907) 1:266–69; Claire Richter Sherman, "Representations of Charles V of France (1338–1380) as a Wise Ruler," *Medievalia et Humanistica* 2 (1971): 86–93; the two exhibition catalogs, *La Librairie de Charles V* (Paris: Bibliothèque Nationale, 1968), 115 and *Les Fastes du gothique, le siècle de Charles V* (Paris: Editions de la Réunion des musées nationaux, 1981), 335–36; North, *Universe*, 44–45; and Laird, *Manuscripta* 34 (1990): 167–76. It will be discussed in Sherman's forthcoming book on verbal and visual representation in fourteenth-century France, projected for 1994.

[81] *Librairie de Charles V*, 115, and *Fastes du gothique*, 335.

[82] Reproduced in Sherman, *Medievalia et Humanistica* 2 (1971): fig. 4, and *Fastes du gothique*, 335.

[83] A miniature from another copy of the *Traité*, in Paris, Bibliothèque Nationale, ms. français 1350, shows Oresme demonstrating the armilliary to a group of four young students. That miniature is reproduced in Cora E. Lutz, *Essays on Manuscripts*

At the beginning of Pèlerin's *Livret* (fol. 33r) is a miniature showing the same crowned figure as at fol. 1r (Charles), seated in the same study among his books, although the armilliary sphere, appropriately to the change in subject matter of the text, has disappeared.[84] A second figure, kneeling, presents a book to the king. This second figure, much younger-looking than the figure representing Oresme at fol. 2r, is presumably Pèlerin.

There is no portrait accompanying Pèlerin's *Practique*, nor are there any technical diagrams with it, although well-executed diagrams, some illuminated, accompany Oresme's *Traité*.[85]

At the beginning of the treatise by Alchabitius (fol. 119r) is a portrait of a single dark-haired figure seated at a desk like that in the royal pictures with a book open before him. His chin rests in the palm of his hand, and he frowns as if in concentration. An ink-pot and quill are beside him. Presumably he represents Alchabitius or his translator.

The two royal miniatures, and probably the other miniatures as well, are the work of an artist who also did miniatures for the magnificent Bible that Jean de Vaudetar presented to Charles V in 1372.[86] That artist was an outstanding close disciple of the master artist of the Bible of Jean de Sy, made for Charles's father, King John.[87]

The royal horoscopes at the end of the manuscript are finely illuminated.[88] That of Charles V (fol. 158v), lettered in gold, is contained in a square decorated with alternating Ks and crowns.[89] That of Charles the dauphin (fol. 159r) is decorated with dolphins and fleurs-de-lis.

---

and Rare Books (New York: Archon Books, 1975), 55.

[84] Reproduced in Sherman, *Medievalia et Humanistica* 2 (1971): fig. 5.

[85] One of these is reproduced in Edgar Laird, "Robert Grosseteste, Albumasar, and Medieval Tidal Theory," *Isis* 81 (1990): 693.

[86] *Fastes du gothique*, 335.

[87] *Fastes du gothique*, 325-26 and 331-32.

[88] These horoscopes are not unique. Simon de Phares reports that the ones made for Charles's children by André de Sully were also very fine ("en moult beau stille"-*Recueil*, 332). See also Emmanuel Poulle, "Horoscopes princiers des XIVe et XVe siècles," *Bulletin de la Société Nationale des Antiquaires de France* (1969), 63-69; and J. D. North, *Horoscopes and History* (London: Warburg Institute, 1986), wherein he says (141) that the French court was more active than the English in producing horoscopes.

[89] Reproduced in *Librairie de Charles V*, pl. 5, and *Fastes du gothique*, 335. It is from this horoscope, sometimes assumed to have been calculated by Pèlerin, that historians derive the date of Charles's birth, Jan. 21, 1338. See Delachenal, "La Date de la naissance de Charles V," *Bibliothèque de l'Ecole des Chartes* 64 (1903): 94-98, and *Histoire* 1:1-2; also Christine de Pisan, *Livre des fais et bonnes meurs du sage roy Charles V*, ed. Solente, 15, and n. 1.

These portraits and horoscopes, along with uncertainty about when Oresme wrote his *Traité*, create some problems in dating the manuscript. The volume, which has been described as "lavishly illuminated," is handsome enough to suggest that it is an original presentation copy.[90] But Pèlerin's treatises were completed in 1361 and 1362, before Charles V became king (1364), so that the crowns in his portraits would have been inappropriate at those dates. Furthermore one of the royal children for whom a horoscope is provided was not born until 1373, so the volume as it now stands can be no earlier than that. The value of these reflections is rendered somewhat doubtful, however, by the likelihood that the horoscopes are later additions and the suggestion that the crowns were painted in when those additions were made.[91]

Two arguments, taken together, imply that the volume could not have been produced either before or after 1377. Unfortunately both arguments are somewhat flawed. The first asserts that one of the royal horoscopes pertains to a child who died in 1377, after which time it is unlikely that such a horoscope would have been calculated. This argument holds, of course, only if the horoscopes are part of the original manuscript. The second argument states that since Oresme's *Traité* was written in 1377, a manuscript containing it could exist no earlier. Authorities on Oresme's canon and chronology, however, are less precise about the date of the *Traité*, placing it somewhere between 1366 and 1377.[92] On the whole, it seems likely that the manuscript was produced sometime around 1377, and more likely earlier than later.[93]

Text and Translation

In preparing our text of the *Practique*, we have for the most part simply transcribed the only known copy of it, adding modern punctuation and diacritical marks and silently expanding standard abbreviations, of which there are comparatively few. All other alterations are

---

[90] North, *Universe*, 44.

[91] Sherman, *Medievalia et Humanistica* 2 (1971): 87; Willard, *Christine de Pizan*, 22.

[92] Menut, *Mediaeval Studies* 28 (1966): 291 and 293, points out that Oresme's *Du Ciel et du monde* (1377) refers to the *Traité* as already in existence and that the *Traité* likewise refers to the *Livre de divinacion* (tentatively, 1366).

[93] The horoscopes on the initial, unnumbered pages are scrawled on the manuscript rather than being produced as part of it. One is dated astronomically at 1384, but it is surely irrelevant to dating the manuscript. See North, *Universe*, 45 n. 10.

noted in the "Textual Notes," which are indicated in the text by superscript letters. Superscript numbers in the text refer to the "Explanatory Notes."

In the translation our goal is to present a clear and readable English text that accurately reflects the intellectual content of the French work. While attempting to stay as close as possible to the original, we have nevertheless simplified some of Pèlerin's more convoluted expressions and broken down some of his lengthy sentences into separate, more comprehensible English sentences. We have also taken the liberty of omitting redundant words in some places, and we have added words in a few other places to clarify the author's meaning. Our translation is not intended to capture the style of the medieval writer but to reproduce his ideas in straightforward English.

Practique de astralabe

[Prologue]

La science du firmament et du mouvement des estoiles en la partie de practique[1] ne puet en aucune maniere bonnement sans instrumens estre mis parfaitement en oeuvre. Pour ce ont les professeurs et executeurs du dit art et science ordené et fabriqué pluseurs instrumens, aucunes en figure reonde[2] et les autres en figure plate,[3] des queles le plus bon et vray et plus acoustumé en usage est ce qui est appelé astralabe, le quel le tresreverent Ptholomee[4] a demonstré par drois commencemens de art de mesure et de nombre estre figuré de tous cercles essencieles et accidenteles[5] imaginés[6] en la figure de la neuviesme spere exstant toute reonde compasser et figurer sans faillir ou dit instrument en plainne et plate figure. Et en ce instrument sont pluseurs proffis et nobles consideracions que un home, comment que il soit bien duit et enformé en partie de practique des mouvemens des estoiles, ne puet retenir tout en memoire.

Pour ce ai je par commandement de mon tresredoubté seigneur, tres haut et noble prince,[7] les proffis et generalz observances communement cheans en practique, pour commun proffit, de la dicte science mis en ce livret en langue françoise[8] afin que chascun, combien qu'il ne soient grans entendeurs de livres en latin sur cé ordenes,[9] s'en puissent aider en aucune maniere.

Et ces profis du dit instrument ai je parti[10] et mis brifment et plus simplement a tout mon povoir en 20 parties ou chapitres, les quiex, s'il sont bien entendus, chascune personne pouvra pluseurs autres proffis de lui meisme ordener et entendre.

Et je pri dieu que il m'adresce a ce parfaire en la voie de verité et perseverance.

Le premier proffit a practique ou chapitre de astralabe est a savoir a chascun jour de l'an en quel signe et degré est le soleil.

Le secont pour prendre la hauteur du soleil et des estoiles.

Le 3 pour trouver les heures de jours et de nuit et le signe ascendent et les 4 angles.

Le 4 a savoir la quantité du temps de soleil couchant jusques a la nuit serree, et du point du jour jusques au soleil levant.

Le 5 de savoir la quantité du jour et de la nuit.

Prologue

In its practical aspect, the science of the firmament and of the movement of stars cannot be accurately undertaken in any way without instruments. For this reason teachers and practitioners of this art and science have designed and made several instruments, some spherical and others flat, of which the most accurate and most widely used is called an astrolabe. The honorable Ptolemy demonstrated by means of the fundamental principles of measurement and numbers that the astrolabe consists of all the essential and accidental circles imagined to exist in the ninth sphere, which is spherical but which is represented on the instrument without distortion as a plane and flat figure. There are so many principles and noble considerations associated with this instrument that a man, no matter how well informed he may be or how much practical knowledge he may have of the movement of the stars, cannot remember everything.

For this reason, I, by order of my redoubtable lord, the very high and noble prince, have put into French in this little book the uses and general rules commonly falling within the general practice of the aforementioned science for the common good. Thus everyone, although he may not understand a great deal from books written in Latin of this sort, can profit in some way from this book.

I have divided up these uses for the astrolabe into 20 parts or chapters and have done my best to describe them briefly and simply. If they are well understood, they will enable each person to understand and derive other practices for himself.

I pray to God that He give me the ability to accomplish this task in the path of truth and perseverance.

The first practical use of the astrolabe or the first chapter is on how to determine which sign and degree the sun is in for each day of the year.

The second, to take the altitude of the sun and the stars.

The third, to find the hours of the day and night, the ascendant, and the four angles.

The fourth, to know the amount of time from sunset to the dark of night and from first daylight to sunrise.

The fifth, to know the length of day and night.

Le 6 pour trouver les heures equales.

Le 7 pour trouver la quantité de l'eure inequale et quele partie de l'eure est passee.

Le 8 a savoir la hauteur du soleil a chascun midi.

Le 9 pour savoir avecques quel degré chascune estoile fixe vient a midi ou avecques quel degré ele se lieve.

Le 10 a savoir en quel endroit des quartes du monde chascune estoile soit a chascune heure assise.

Le 11 pour savoir a combien de temps chascun signe se lieve ou couche.

Le 12 pour cognoistre les estoiles fixes ou firmament dedenz le astralabe assises.

Le 13 pour trouver le vray midi en chascune vile.

Le 14 a savoir trouver par le astralabe le signe de la lune et des estoiles.

Le 15 pour trouver la latitude des planetes et les retrogradacions.

Le 16 pour trouver les 12 maisons.

Le 17 a savoir la hauteur du soleil et des estoiles par les heures.

Le 18 pour mesurer la hauteur des choses.

Le 19 pour mesurer le lonc d'un champ.

Le 20 pour savoir la hauteur des choses par leur ombres faites du soleil et de la lune.

[Part I]

Devant les proffis devant dis faut il monstrer et exposer et savoir nommer les cercles et lignes de astralabe.

[1.] C'est assavoir que le instrument appelé "astralabe"[11] est ront en sa figure plate[12] et figuré a chascun des costés de divers tres, cercles, et figures. Et l'une parcelle ou il a une roe[13] qui se tourne est appellé "visage" ou "face."[14] Et l'autre part est dicte le "dos."[15]

[2.] Et tout le instrument est ront comme dit est, fors que en 1 lieu ou il a une hautesse petite.[16]

[3.] Et est illec atachié un annel nommé "armille"[17] qui se tourne, par le quel annel le instrument nous devons prendre a nostre doit quant mestier est.

[4.] Apres est appelee "mere"[18] toute la reondesse et le font qui comprent les tables[19] devers le visage de l'instrument.

[5.] Et entour le tous est atachié un cercle reont espés, nommé le "lymbe"[20] qui comprent les tables, et est parti en 360 parties.

The sixth, to find the equal hours.

The seventh, to find the length of unequal hours and how much of an hour has passed.

The eighth, to know the altitude of the sun at noon each day.

The ninth, to know with which degree each fixed star arrives at the meridian or with which degree it rises.

The tenth, to know in which direction, with respect to the four quarters of the world, each star is located at each hour.

The eleventh, to know how much time it takes for each sign to arise or to set.

The twelfth, to identify the fixed stars of the firmament which are marked on the astrolabe.

The thirteenth, to find the true meridian in each town.

The fourteenth, to know how to find the sign of the moon and the stars by means of the astrolabe.

The fifteenth, to find the latitude of the planets and the retrogradations.

The sixteenth, to find the twelve houses.

The seventeenth, to know the altitude of the sun and the stars by knowing the hour.

The eighteenth, to measure the height of things.

The nineteenth, to measure the length of a field.

The twentieth, to know the height of things by the shadows they cast by the sun and the moon.

[Part I]

Before going on to the above-mentioned uses, it is necessary to identify, to explain, and to name the circles and lines of the astrolabe.

[1.] We must know that the instrument called an "astrolabe" is round and flat and is engraved on each side with various lines, circles, and figures. The side on which the movable wheel is located is called the "visage" or the "face." The other side is called the "back."

[2.] The whole instrument is round except in one place where it has a small raised area.

[3.] Here is attached a movable suspending ring called an "armilla." When we use the astrolabe, we hold it by this ring on our finger.

[4.] The round receptacle which holds the plates on the front of the instrument is called the "mother."

[5.] Around the whole instrument, a broad circular band called the "limb" surrounds the plates and is divided into 360 parts.

[6.] Chascune des tables illec encloses tient 3 cercles des quiex le plus grant est appelé "cercle de Capricorne," le moien le "equinoctial,"[21] et le mendre "du Cancre."[22]

[7.] Et sont les tables trenchiees de 2 trais drois a guise d'une crois parmi. Et le bras de la crois venant de annel ou chascune table a un dentelet par le quel elle se tient a un trou dessoubz l'annel, le dit trait jusques a milieu de la table est appelé "lingne de midy." Et l'autre mutet du milieu de la table jusques au bort est appelé "lingne de minuit."[23] Et l'autre trait travers lonc parmi la table[24] est nommé "lingne de vrai orison" ou "vray orient et occident."

[8.] Apres a il pluseurs cercles figurés sur les tables, des queles a une maniere qui sont un greigneur de l'autre, et aucuns sont entiers et tous reondes, et les greigneurs ne sont pas entieres. Et tous ces cercles de ceste maniere sont appelés "almucantharat" en langue de Arabie.[25] Et le greigneur de tous est nommé "orison oblique,"[26] et il signifie le cercle du ciel qui depart la muté visible de la invisible.[27]

[9.] Et quant nous regardons le astralabe devers l'annel tenant le instrument sur le plat d'entre nos 2 mains toute la muté de la lingne de midi et de minuit devers le pote main est appelee "orientele" et par consequent les parties des cercles illec figurees sont appelés "orien-[te]les." Et l'autre moitié est "occidentele."[28]

[10.] Le centre ou le milieu du plus mendre almucantharat est dit "cenith"[29] et signifie le point du firmament qui est sur nostre chief droitement assis.

[11.] C'est assavoir que chascune table de chascun astralabe doit avoir 90 almucantharat, mais il faut pour ce faire le instrument trop grant, au mains de largeur d'un pié. Et ainsi les instrumens fussent trop pesans et non portatis.[30] Et pour ce les oeuvrers font aucune fois almucantharat pour 2 ou pour 3 ou un pour 4 ou pour 5, selonc la grandeur de l'instrument. Et quant la table tient 45 almucantharat chascun vault 2. Et se il tient 30 chascun vault 3. Et se la table tient 18 chascun vault 5. Et s'il tient 15 chascun est de 6.

[12.] Apres a il les "azimuth"[31] qui sont cercles grandes qui se trenchent tous par le cenith et tous ne sont pas entiers, car il ne passent mie l'orison. Et doivent estre a un parfait astralabe[32] 90 a chascune quarte partie de la table, c'est a dire d'entre midi et vray orient ou midi et occident vraye.

[13.] Et dessoubz le orison a il 12 traiz avecques la lingne de minuit[33] qui signifient les heures inequales.[34]

[14.] Cause pour quoy il faut faire pluseurs tables a chascun astralabe est celle que les jours et les nuis en un pais sont plus longues que

[6.] Each of the plates enclosed therein holds three circles of which the largest is called the circle of Capricorn, the middle one the equinoctial circle, and the smallest one the circle of Cancer.

[7.] The tables are engraved with two straight lines forming a cross in the middle of the plate. One arm of the cross extends from the suspending ring to a hole below it [in the center of the astrolabe] where the plates are attached by a pin; and that arm is called the "line of midday." The other arm from the center of the plate to the rim is called the "line of midnight." The other line crossing the plate horizontally is called the "line of true horizon" or "true east and west."

[8.] There are several circles engraved on the plates, some of which are larger than the others. Some of them are whole circles, while others—the larger ones—are partial circles. All the circles of this kind are called "almucantars" in Arabic. The largest of all these circles is called the "oblique horizon." It represents the circle which divides the visible half of the sky from the invisible half.

[9.] When we hold the astrolabe flat between our hands, looking [across it] toward the suspending ring, the half from the line of midday and midnight toward our left is called "oriental," and consequently the parts of the circles contained therein are called oriental. The other half is "occidental."

[10.] The center of the smallest almucantar is called the "zenith" and represents the point of the firmament which is located directly over our head.

[11.] We must know that each plate of each astrolabe should have 90 almucantars, but if it did, the instrument would be too big, at least a foot wide. The instrument would be too heavy and non portable. Hence, craftsmen sometimes mark every second, third, fourth, or fifth almucantar, according to the size of the instrument. When the plate has 45 almucantars, each one is worth two. If it has 30, each one is worth three. If the plate contains 18, each one is worth five, and if it has 15, each one is worth six.

[12.] Next, the plate has "azimuths" which are great circles, all of which pass through the zenith and are not complete circles because they never go beyond the horizon. In a perfect astrolabe there should be 90 degrees in each quadrant, that is, between the meridian and true east or the meridian and true west.

[13.] Beneath the horizon, there are 12 lines, which together with the line of midnight represent the unequal hours.

[14.] The reason it is necessary to make several plates for each astrolabe is that the days and nights in one country are longer than in

en un autre.³⁵ Et pour savoir et cognoistre la table par quoy nous devons ouvrer^a en chascun pays—pour ce cognoistre nous fault savoir la hauteur de l'estoile marine³⁶ a chascun pais et compter de orison les almucantharat jusques au milieu de la table ou du equinoctial jusques a cenith. Et celle que nous trouverons plus pres de nostre nombre, cele doit estre mise sur toutes autres, et devons ouvrer par elle ou pays ou nous sommes.

[15.] Pardessus les tables a il une roe toute persee dite le "reys" ou "rethe,"[b] ³⁷et elle tient figure le zodiaque large ou il a 12 parties ou sont escris les 12 signes, et chascun signe est partie en 30 degrés ou en 15 mais donques un vault 2. La circumferance de zodiaque ou les degrés sont est appelé "ecliptique," et le reys est quartilé de 2 traiz travers ainsi comme les tables, et chascun quart comprent 3 signes. Et toute la partie du dit reys enclos d'un cercle³⁸ qui se fait en tournant le reys du commencement du chief de Aries et de la Libre est appelee "septentrionele." Et tout ce qui est dehors le dit cercle est appelé "meridionel."³⁹ En autre maniere est appelé tout septentrionele ce qui est enclos du zodiaque, et ce qui est dehors est certaine meridionel. Les bous dedenz longuetes faites en pluseurs manieres a pluseurs lieus de rethe signifient les estoiles fixes du firmament plus grans et notables. Et lors nons est escript sur chascun dent, comme Alhabor, Altaire, Aldebaran⁴⁰, et ainsi des autres.

[16.] Au commencement de Capricorne⁴¹ est un petit dent qui touche au bort de la mere qui est clamé "almuri," c'est a dire le monstreur.⁴²

[17.] Et depuis a il un trou ou milieu de reys et des tables a la mere, au quel est une cheville qui tient tout ensemble, qui puet estre appelé "cheval"⁴³ ou autrement a nostre plaisir.

[18.] En l'autre partie de l'instrument qui est devant nomé le dos a il 2 manieres de cercles⁴⁴ des quiex celi dehors tient les 12 signes et leur degré. Et par un autre nombre tient chascune quarte du dit cercle 90 degrés pour prendre la hauteur des estoiles. L'autre compas tient 12 moys et le jour de l'an.

[19.] Et en la basse partie par dedens a il un demy quarré⁴⁵ de quoy les costés sont partis en 12 et vault pour mesurer la hauteur et longueur des choses. Et sur le quarré aucun astralabe tient cercles qui s'assemblent parmi le trou de l'instrument⁴⁶ et sont appelés "cercles des heures."

[20.] Derrenierement a il la reigle⁴⁷ qui se tourne, et a chascun bout un petit tablet drecié qui a un trou ou 2 pour prendre la hauteur du soleil et des estoiles.

another one. To know which plate to use in each country, we must know the altitude of the north star in each country and count the almucantars from the horizon to the middle of the plate or from the equinoctial line to the zenith. The plate we find nearest to our number is to be placed on top of the others, and we must use this plate in the country where we are.

[15.] Over the plates there is a disc with holes in it called the "rete." This disc represents the broad zodiac in which there are 12 parts with the 12 signs written in them. Each sign is divided into 30 degrees or into 15 parts in which each part is worth 2 degrees. The circumference of the zodiac where the degrees are marked is called "ecliptic," and the rete is quartered by two lines across it, just as the plates are. Each quarter contains three signs. All that part of the rete which is enclosed in a circle traced by the heads of Aries and Libra when the rete is turned is called "northern." All that is outside of this circle is called "southern." According to a different manner, whatever is enclosed in the zodiac is called northern, and whatever is outside is clearly southern. The tips of the little pointers which have different forms at different places on the rete represent the more notable fixed stars of the firmament. Their names are written on each tooth, such as Alhabor, Altaire, Aldebaran, and so on.

[16.] At the beginning of Capricorn there is a small tooth which touches the rim of the mother and which is called "almuri," that is, the indicator.

[17.] In a hole in the middle of the rete and the plates of the mother there is a peg which holds everything together, which we can call a "horse" or any other name we wish.

[18.] In the other part of the instrument which we earlier called the back, there are two kinds of circles of which the outer one holds the 12 signs and their degrees. Each quadrant of this circle is also numbered from zero to 90 degrees for taking the altitude of the stars. The other circle shows the 12 months and the days of the year.

[19.] In the lower part within the circle there is a half square whose sides are divided into 12 parts which are used to measure the altitude and distance of things. Above the square, some astrolabes have circles which come together in the hole of the instrument and which are called "circles of the hours."

[20.] Finally, there is a movable rule. Each end of it has a small upright vane containing one or two holes for taking the altitude of the sun and the stars.

[Part II]

1. *Pour savoir a chascun jour de l'an en quel signe et degré est le soleil.*⁴⁸

Nous devons sus le︎ᶜ dos de l'instrument⁴⁹ tourner la riulle sur le moys au quel nous sommes, et sur la presente journee, et apres regarder quel signe et degré au cercle des signes touche le bout de la riulle; car en tel est le soleil en firmament ou assés pres de lui. Toutefoiz il est plus seure chose prendre le lieu du soleil⁵⁰ par le almanach.⁵¹

Quant nous savons le lieu du soleil par estimacion nous pourrons trouver les jours du moys, car nous tournerons la riulle sur le degré du soleil et elle trenchera le moys et le jour au quel nous sommes. Et a semblabe maniere⁵² comme il est dit devant porrons nous trouver le lieu du soleil sur quelconques journee presente ou autre.

2. *Pour savoir la hauteur du soleil par jour et des estoiles par nuit.*⁵³

Nous devons prendre le astralabe par le annel franchement a un de noz doiz⁵⁴ et tourner le costé devers le soleil et lever et baissier⁵⁵ la riulle tant de foiz que lé rais du soleil venant par la tablete sur le bout de la riulle dresciee ⟨par le trou soit contrepondent contre l'autre tro de la table a l'autre bout de la riulle dresciee.⟩ᵈ Et donques devons nous regarder combien la riulle est remuee de la lingne de vray orient et occident par les divisions et nombre, et telle sera la hauteur du soleil a l'eure de notre consideracion. Aucune foiz avient que le soleil ne fait point de clarté, toutesfois le cours⁵⁶ appert parmi les raies. Et donques convient regarder de l'ueil parmi les 2 tros ⟨et la riulle tant de fois lever et bessier jusques que nous voions le soleil parmi les 4 tros.⟩ᵉ⁵⁷

Et ne plus ne mains devons faire par nuit a savoir la hauteur des estoiles fixes ou des planetes. Et devons estre bien cauteleus a ce faire, car les estoiles ne pueent aucune foiz bien apparoit[re] parmi les tros, et doncques nous le prenons par droiture de l'une. Et pour ce est bon que les tabletes aient 2 paires de tros,⁵⁸ une pere greigneur que l'autre, ou 2 petis dentelés,⁵⁹ l'un contre l'autre ou milieu au bout des tabletes.

3. *Quant nous voulons savoir quelle heure il est des jours et de nuit et le signe et degré ascendent par le astralabe.*⁶⁰

C'est assavoir que il a 2 manieres de heures des queles les unes sont appelees equales et les autres inequales. La equale heure est la 24ᵉ partie de jours et de la nuit tout ensemble. Mais inequale est la 12ᵉ partie de jours ou de la nuit, se lé jours ou les nuis sont lons ou courtes. En ce chapitre doit estre entendu des inequales heures.

[Part II]

1. *How to determine for each day of the year the sign and degree the sun.*

On the back of the astrolabe we have to turn the rule to the current month and day and note which sign and degree on the circle the end of the rule touches, for the sun in the firmament is located at this position or quite near to it. However, it is more accurate to find the position of the sun from an almanac.

When we know the approximate position of the sun, we will be able to find the days of the month, because as we place the rule on the degree of the sun, it will lie across the current month and day. In a similar manner we can find the place of the sun on any day whatsoever.

2. *How to take the altitude of the sun by day and the stars by night.*

We have to hold the astrolabe firmly by the ring by one of our fingers and turn the edge toward the sun. We raise and lower the rule until the rays of the sun coming through the vane on the end of the rule thus set, pass through the corresponding hole of the vane on the other end of the rule. We have to take note of how many numbered divisions the rule has been moved from the line of true east and west. This difference will be the altitude of the sun at the time we are taking the measurement. Sometimes it happens that the sun is not shining, but we can still locate the sun by its rays. In this case, it is good idea to peer through the two holes and raise and lower the rule until we see the sun through both holes.

We do the same thing at night to determine the altitude of the fixed stars and the planets. We have to be careful doing so, because sometimes the stars cannot be clearly seen through both holes. Therefore, we take it by the alignment of the one. For this reason it is good that the vanes have two pairs of holes, one pair larger than the other, or two little teeth, one opposite the other centered on top of each vane.

3. *When we want to know the time of day and night and the ascendant degree by means of the astrolabe.*

We must know that there are two kinds of hours, of which one is called equal and the other unequal. An equal hour is the 24th part of the day and night taken together. An unequal hour is the 12th part of the day or of the night, regardless of whether the days or nights are short or long. In this chapter we are talking about unequal hours.

First we have to find the position of the sun as shown in the first chapter. Then we have to take the altitude of the sun, if it is day time,

Nous devons premierement a ce faire trouver le lieu du soleil comme le premier chapitre monstre, et apres devons prendre la hauteur du soleil se il est de jours ou la hauteur d'une estoile fixe se il est par nuit par la maniere du secont chapitre. Et devons considerer se le soleil ou l'estoile est d'entre orient et midi ou d'entre midi et occident.

Et quant ce est fait donques nous devons figurer le soleil et son nadair, c'est a dire son opposit[61] sur la ecliptique, sur le zodiaque; et tourner le zodiaque autant que le lieu du soleil soit sur cel nombre de almocantharat, le quel nous avons pris par la hauteur du soleil d'entre orient et midi et d'entre midi et occident. Et doncques le opposit lieu du soleil d'entre les trais des heures monstre l'eure de jours inequale; et le signe et degré du zodiaque qui touche le orison devers orient est montant, ou ascendent, a l'eure; et le signe qui touche la lingne de midi et qui touche le orison occidentel et la minuit commencent les autres quartes.

Et s'il nous plait a savoir l'eure par nuit, le bout du dent de l'estoile fixe la quele nous avons pris par la hauteur devons mectre sur le almucantharat semblable, ainsi comme nous avons dit du soleil. Et doncques le lieu du soleil monstrera l'eure, et le ascendent monstrera le signe qui cherra sur le orison devers orient.

Et se le lieu du soleil ou le lieu de l'estoile par la hauteur chiet d'entre 2 almucantharat, doncques nous devons mectre le lieu du soleil ou l'estoile fixe sur le prochain almucantharat plus bas[f] et noter le almuri au droit du bor[t] de la mere, et apres tourner le rete jusques le lieu du soleil ou de l'estoile chiee sur le prochain almucantharat plus haut, et noter combien le almuri est remué. Et de ce devons prendre telle partie selonc ce que la hauteur du soleil ou de l'estoile soit plus grant du premier almucantharat, et sur ce mectre le almuri sur le bort et doncques cherra le soleil ou l'estoile d'entre les 2 almucantharat droitement sans deffaute.

C'est a noter que nulle heure du jour est tant a doubter[62] de prendre l'eure ou le ascendent comme pres de midi, quer le soleil tient comme une hauteur, et il est doubté se il est devant midi ou apres. Et pour ce est tres bonne et convenable chose savoir le midi par l'ombre du soleil en chascun lieu ou nous convient de morer. Et de ce pensai je a parler plus plainement ci-apres ou il sera a propos ou 13e chapitre.

4. *Se nous voulons savoir l'espace ou temps du soleil couchant jusques a la nuit serree ou du point du jour jusques a soleil levant.*[63]

Nous devons pour le vespertin crepuscle[64] mectre le soleil sur l'orison occidentel et noter le almuri et tourner le rethe a tant que le na-

or the altitude of a fixed star, if it is night, as shown in the second chapter. Next we have to consider whether the sun or the star is between the east horizon and the meridian or between the meridian and the west horizon.

When this is done, we set the sun and its nadir (that is, its opposite on the ecliptic) on the zodiac. To do so, we turn the zodiac so that the position of the sun is on the almucantar representing the altitude we found for the sun between the east horizon and the meridian or between the meridian and the west horizon. Then the place where the nadir of the sun falls among the lines of the hours shows the unequal hour of the day. The sign and degree of the zodiac which touches the horizon towards the east is rising, or ascending, at that time, and the signs which touch the meridian, the western horizon, and the line of midnight begin the other quarters of the zodiac.

If we want to know the time at night, we have to place the end of the tooth representing the fixed star whose altitude we have taken on the corresponding almucantar, just as we did for the sun. Then the position of the sun will show the hour, and the ascendant will show the sign which falls on the horizon towards the east.

If the altitude of the sun or the star falls between two almucantars, we have to put the longitude of the sun or the fixed star on the next lower almucantar and note where the almuri falls on the border of the mother. Then we turn the rete until the longitude of the sun or the star falls on the next higher almucantar and note how much the almuri has moved. We must calculate the proportion which the altitude we have taken bears to the nearest almucantar on the astrolabe, and by that same proportion we set the almuri between its previously noted settings in the border. The sun or the star will assuredly fall between the two almucantars.

It should be noted that no time of day is more uncertain for telling the time or finding the ascendant than around noon, because the sun stays at about the same altitude, and it is hard to tell whether it is before or after noon. For this reason it is much better to find the noon hour by the shadow of the sun in the location where we happen to be. I intend to discuss this subject more fully in chapter 13, where it is more appropriate.

4. If we want to know how long it is from sunset to the dark of night and from first daylight to sunrise.

To find the evening crepuscule, we have to place the longitude of the sun on the western horizon and take note of the almuri. We then

dir du soleil[65] soit haussé par 18 almucantharat[66] et par tant durra la clarté comme le almuri se change des degrés de limbe. Et le point du jour commence quant le nadir du soleil vient sur le 18 almucantharat devers occident.

5. *Pour savoir la quantité du jour et de la nuit.*[67]

Ce endroit est l'eure entendue du jour et de la nuit artificiele,[68] c'est a dire, du soleil levant jusques au couchant soleil, et la nuit de couchant jusques a levant. En ce cas devons mectre le lieu du soleil sur l'orison orientel et noter le almuri au droit du limbe[69] et tourner le rethe jusques a ce que le degré du soleil chiee sur l'orizon occidental et illec aussi noter le lieu de almuri. Doncques, ce que le almuri a passé du limbe est la quanité de jour, et le remenant est la quanité de nuit.

Et est assavoir que tous les 2 jours sont equalz et semblabes quant le soleil se tient egalment du commencement du Cancre ou de Capricorne, et par consequant leur nuis sont equales. Et pour ce est il vrai que les degrés opposites ont leur nuis et leur jours equales un a autre. Car si lonc est le jour existant le soleil au commencement du Cancre comme la nuit est existant le soleil en Capricorne.[70]

Et en semblabe maniere[71] devons faire des estoiles fixes et des planetes pour savoir combien il demeurent dessus ou dessoubz terre.

6. *Pour savoir combien des heures equales sont passees apres soleil levant ou apres midi.*[72]

Nous devons mectre le soleil sur sa hauteur d'entre les almucantharat ainsi comme le tiers chapitre nous ensaingne, et illec devons noter le almuri. Et devons remener le lieu du soleil devers orient ou sur le midi; et tant de fois que le almuri passe 15 degrés du limbe,[73] tant de heures equales a il jusques a nostre consideracion.

Et se il a avecques ce 5 degrés ou 6 ou quelconques nombre mendre de 15, pour ce devons prendre tele partie de heure comme le dit nombre se a devers 15; et pour 5 degrés font une tierce partie d'une heure et 9 font trois cinquiesmes parties d'une heure equale.

7. *Pour savoir la quantité d'une heure inequale par jour.*[74]

Le opposit degré du soleil devons mectre sur le commencement d'une des heures inequales et noter le almuri et bougier apres le rethe jusques a tant que le opposit viengne sur le bout de l'eure. Et tant comme le almuri se longe, tant est la quantité [de l'] heure journele. Et se nous faisons en telle maniere du degré du soleil nous trouverons la quantité des heures nocturne.

have to turn the rete until the nadir of the sun is raised by 18 almucantars; and by as many degrees as the almuri moves in the border, by that much the light will last. Day break begins when the nadir of the sun arrives at the 18th almucantar towards the west.

5. *To find the length of day and night.*

In this section an hour is understood to be a division of artificial day (that is, from sunrise to sunset) and of artificial night (from sunset to sunrise). In this case we have to place the longitude of the sun on the east horizon and note where the almuri points on the rim. Then we turn the rete until the degree of the sun falls on the west horizon and note the place of the almuri there. Thus the number of degrees the almuri has passed on the border represents the length of day, and the remainder is the length of the night.

We must know that on any two days when the sun is equally distant from the beginning of Cancer or of Capricorn, those days are of equal length, and consequently so are their nights. For, the length of the day when the sun is at the beginning of Cancer is equal to the length of the night when the sun is at the beginning of Capricorn.

We work in a similar way with the fixed stars and the planets to know how long they remain above or below the earth.

6. *To find how many equal hours have passed since sunrise or since noon.*

We have to place the longitude of the sun on the almucantar of its altitude, just as the third chapter taught us, and there take note of the almuri. We then bring the longitude of the the sun back to the east horizon or the meridian; the number of times the almuri traverses 15 degrees on the limb is the number of equal hours up to the time we are considering.

If there are in this result 5 degrees or 6 or some number less than 15, then we have to take that fraction of the hour which that number is to 15; thus, 5 degrees makes a third of an equal hour and 9 makes three-fifths of an equal hour.

7. *To find the length of an unequal hour by day.*

We must place the nadir of the sun on the beginning of one of the unequal hours and note the almuri. We then move the rete enough so that the nadir falls on the end of the hour. The distance that the almuri has moved is the length of the hour of the day. If we work in the same way using the longitude of the sun, we will find the length of the hour of the night.

Et se il est passé une partie de l'eure du soleil et nous voulons savoir combien il est passé d'elle, nous devons signer le lieu de almuri et remener le opposit du soleil par jour ou le lieu du soleil par nuit sur le commencement de l'eure et regarder quelle partie est le nombre que le almuri a fait. Devers la toute heure, tele partie est passé de l'eure.

8. *Pour savoir la hauteur du soleil a chascun midi ou de quelconques lieu du zodiaque ou des estoiles fixes.*[75]

Pour ce devons mectre le lieu du soleil, ou de quelconques signe nous voulons ce faire, ou le bout de l'estoile fixe sur la lingne de midi. Et devons compter de l'orison toutes les almucantharat jusques a la lingne de midi ou le soleil ou l'estoile fixe reposé; et quel nombre nous trouverons, tele sera la hauteur de midi ou du lieu du zodiaque ou de l'estoile fixe.

Quant nous savons la hauteur de midi[76] ou du lieu du zodiaque ou de l'estoile et voulons trouver les heures inequales par la alidada, c'est la riulle sur le dos de instrument, donques devon mectre la riulle sur la hauteur meredionele et garder en quel endroit la riulle le touche, le entier compas des heures sur le dos figurees.[77] Et illec devons figurer la riulle. Et toute la journee, quant il nous plaira, nous prendon la hauteur du soleil et regarderons en quel endroit le signe fait en la riulle soit entre les heures. Et c'est droitement tel ouvrage de savoir les heures comme par le quadrant,[78] par le fil, et la parle. Toutefoiz devons savoir que ceste ouvrage n'est une si vraye comme celle qui est dite devant ou tiers chapitre de trouver les heures.

9. *Pour savoir quel degré du zodiaque est en midi avecques chascune estoile fixe ou en orient quant ele se lieve.*[79]

Nous devons mectre l'estoile sur la lingne de midi ou sur l'orison orientel ou occidentel, et donques devons considerer quel degré du zodiaque est en midi ou en orient ou en occident; et tel se lieve ou couche ou est en midi avecques la dite estoile.

10. *Pour savoir en quel endroit est le soleil ou la lune ou les autres estoiles d'entre orient et midi ou d'entre midi et occident ou en quel endroit et combien elles se lievent de vray orient.*[80]

Nous devons prendre la hauteur du soleil ou de la lune ou de quelconques estoile nous vouldrons. Et devons celle mectre sur le semblabe almucantharat comme nous faisons pour trouver les heures et le ascendent.[81] Et devons compter du premier azimuth, c'est ce cercle qui touche l'orison en tel lieu ou la lingne li trenche du vray orient et

If part of the hour of the sun has passed and we want to know by how much it has passed, we have to mark the place of the almuri and bring back the nadir of the sun by day or the longitude of the sun by night to the beginning of that hour. Then we compare the number of degrees the almuri has moved to the number of degrees in the whole hour, and that proportion represents the part of the hour that has passed.

8. *To find the altitude of the sun, of any point of the zodiac, or of any fixed star at the meridian.*

We have to put the longitude of the sun, or of any sign we want to measure, or the pointer of the fixed star on the line of midday. We then count all the almucantars from the horizon to the point on the line of midday where the sun or the fixed star rests; the number we find will be the meridional altitude [of the sun], or of the point of the zodiac, or of the fixed star.

When we know the meridional altitude of the sun, of the point of the zodiac, or of the star, and we want to find the unequal hours by the alidade, that is, the rule on the back of the astrolabe, we have to put the rule on the meridional altitude and note where the rule touches the only hour line on the back of the astrolabe that is a complete circle. At that place we must mark the rule. And at any time of the day we please, we take the sun's altitude and see where the sign made on the rule falls among the hour lines. This is exactly how we work with the quadrant to find the hours by string and bead. Nevertheless we must know that this way of finding the time is not as accurate as the one which was described in the third chapter.

9. *To know which degree of the zodiac is on the meridian with each fixed star or on the east horizon with each fixed star.*

We have to put the star on the line of midday or on the eastern or western horizon and then consider which degree of the zodiac is at the meridian, the eastern, or western horizon. That degree rises or sets or reaches the meridian with this star.

10. *To know in which direction the sun, the moon, or the other stars are located between the east and south, or between the south and west, or wherever the point of their rising is located, and how far it is from true east.*

We have to take the altitude of the sun, the moon, or any other star we wish. We have to put this on the corresponding almucantar, as we do to find the time and the ascendant. We have to count from the

occident et passe parmi de cenith. Et combien nous trouverons des azimuth jusques a l'estoile en tel endroit est la dicte estoile assise.

Se nous ainsi metons l'estoile sur l'orison et comptons les azimuth de la lingne orientele jusques a la dicte estoile, nous trouverons en quel endroit ele se lieve, ou plus pres de orient ou de septentrion ou midi.

11. *Pour savoir a combien de temps chascun signe se lieve ou couche en nostre region.*[82]

Pour ce devon mectre le commencement du signe sur l'orison orientel et signer le lieu[83] de almuri sur le lymbe, et devons tourner le rethe a tant que la fin du dit signe touche le orison. Et tant d'espace et degrés que le almuri fait du limbe, a tant de temps se lieve le dit signe en faisant de 15 degrés une heure. Et ne plus ne mains devons faire en occident pour savoir en combien de temps une signe se couche.[84]

Nous devons savoir que les signes 6 qui se lievent avecques mains de 30 degres sont apelés obliques ou de courtes ascencions.[85] Et les autres signes 6 qui se lievent avecques plus de 30 degres sont clamés droites et de longues ascencions.[86] Et chascun signe qui se lieve droitement, le dit signe se couche obliquement.[87]

12. *A savoir cognoistre aucunes des estoiles fixes qui sont mises dedenz le rethe.*[88]

Nous devons considerer aucune estoile fixe de nostre cognoissance et prendre la hauteur de elle; et devons celle mectre sur les almucantharat en la partie ou elle est.[89] Et apres devon regarder les estoiles non cogneues en quel part elles sont assises devers orient et occident et en quele hautesse elles sont d'entre les almucantharat. Et sur celle hauteur devons mectre la riulle sur le dos de astralabe et a celle heure meismes nous tourner devers l'estoile sans bougier la riulle. Et la greigneur estoile qui nous apperra parmi les 2 tros, c'est celle que nous avons consideré, sans nulle deffaute.

13. *Pour savoir le vray midi et le vrai orient en chascune region ou en chascun pays.*[90]

Ce chapitre est molt subtil et grandement neccessaire, especialment pour rectifier horologes, et pour savoir le vray midi, et pour prendre le ascendent entour midi. Et est aussi neccessaire pour trouver le lieu des planetes et pour rectifier les estoiles. Pour ce devons prenre et eslire une place plainne et descouverte ainsi que le soleil toute la journee ou la greigneur partie par toute l'annee puisse illec estre regardee en faisant son ombre. Et quant la place convenable est esleue et faite plainne et equale sans declinacion, doncques prendrons nous la hau-

first azimuth, that is, the circle which passes through the zenith and touches the horizon in the place where the line of true east and west crosses it. The number of azimuths we find to the star shows where that star is located.

If we put the star on the horizon and count the azimuths from the east line to that star, we will find in which place it rises, whether it is closer to the east, the north, or the south.

11. *To know how much time it takes for each sign to rise or to set in our region.*

We have to put the beginning of the sign on the east horizon and mark the place of the almuri in the limb. Then we turn the rete so that the end of this sign touches the horizon. The amount of space, that is, the number of degrees, that the almuri moves in the border is the time in which that sign rises, counting 15 degrees as an hour. We have to do the same in the west to know how much time it takes the sign to set.

We should know that the six signs which rise with fewer than 30 degrees are called oblique, or of short ascension. The other six signs, which rise with more than 30 degrees, are called vertical, or of long ascension. Each sign which rises vertically sets obliquely.

12. *To identify any of the fixed stars that are placed on the rete.*

We have to consider some fixed star we know and take its altitude. Then we put it on the almucantar in the part where it is. Next we observe in which part, towards the east and west, the unknown stars are located and in what altitude they are among the almucantars. We then set the rule on the back of the astrolabe on this altitude and immediately turn towards the star without moving the rule. The biggest star which appears in the two holes is undoubtedly the one that we have been seeking.

13. *To find true noon and true east in each region or in each country.*

This chapter is very subtle and very necessary, especially for setting clocks, knowing the true meridian, and taking the ascendant around the time of noon. It is also necessary for finding the place of planets and for ascertaining the stars. We have to choose a level and open place where the sun can be seen casting a shadow all day or the greater part of the day throughout the year. When the appropriate place has been chosen and made level without any incline, we take the precise altitude of the sun and put it exactly in its place on the almu-

teur du soleil precieusement et la metrons sur les almucantharat droitement en sa partie. Et comptons les azimuth commençant de la lingne de vray orient et tel nombre que nous trouverons de azimuth, sur cela metrons nous la riulle sur le dors. Et prendrons le astralabe entre nos 2 mains par le costé, tournant la face de astralabe devers terre, et faisons le reys du soleil passer parmi noz troes; et doncques nous coucherons le astralabe belement sur terre. Doncques les 4 quartes du astralabe par ces trez principaus partiront le monde en 4 quartes selonc le vray midi, vray orient, occident et minuit. Et de cé quatre pointes du limbe de astralabe nous tirons les trez sur la terre ou sur une pierre ou aucune autre dure chose.

Et ou milieu de la crois nous drecerons un bastonnet reont et par en hault aguet qui puisse faire une ombre par le soleil. Et par toute l'annee et tous jours, quant l'ombre du soleil venra contre le midi devers minuit,[91] il sera vrai midi en celle vile en la quele la besoingne est faite.

14. *Pour trouver le lieu de la lune ou de une autre planete en quel signe il sont et en quel degré.*[92]

Nous devons actendre pour ce tant que le planete soit droitement devers midi, et ce pourrons nous savoir par le chapitre ci-devant dit. Et a celle heure devons prendre la hauteur de une estoile fixe de nostre cognoissance et par elle faire le ascendent aussi comme nous a monstré le 3ᵉ chapitre[g]. Et doncques le degré du zodiaque qui cherra sur midi est le lieu de la lune ou du planete.

Et par jour pourrons nous faire cest ouvrage de la lune quant elle est a midi[93] et par le soleil eu lieu de l'estoile fixe. Et par nuit quant la lune ou le planete ne vient a midi mais appert devers orient ou occident, adoncques nous pourrons ouvrer par contrarie: ainsi que nous entendons que aucune des estoiles fixes soit a midi, et a celle heure nous prenrons la hauteur de la lune ou du planete et ferons la figure[94] par l'estoile fixe. Et puis nous regarderons devers la partie de astralabe ou la lune ou le planete est, et quel degré de zodiaque nous trouverons entre les almucantharat de semblabe hauteur de la lune ou du pla[ne]te, en tel degré est la lune ou le planete.

Et par ceste maniere pourrons nous trouver le lieu des estoiles fixes qui ne sont mie situés en astralabe en prenant la figure du ciel par aucune maniere et les estoiles soient en midi. Mais ceci devons nous sauvement faire entendre, car ce degré que nous ainsi trouverons, c'est celui avecques le quel la dite estoile vient a midi. Mais la declinacion[h] de la dicte estoile est adoncques legierement a congnoistre[95] si nous aviserons les almucantharat d'entre le equinoccial et la hauteur de l'estoile quant elle est droitement a midi.

cantar. We count the azimuths beginning with the line of true east, and whatever number of azimuths we find, we put the rule on that number on the back. We take the astrolabe by the edge between our two hands and turn the face towards the earth so that the rays of the sun pass through our holes. Then we place the astrolabe carefully on the earth. The principal lines of the 4 quadrants of the astrolabe divide the world in 4 quarters according to true south, east, west, and north. From these four points in the border on the astrolabe we draw the lines on the ground, on a rock, or other hard thing.

In the middle of the cross we erect a small round stick with a pointed top which can make a shadow from the sun. Every day in the year, when the shadow of the sun arrives opposite the meridional line on the midnight line, it will be true noon in this town in which this work is done.

14. *To find the sign and degree of the moon or another planet.*

We have to wait until the planet is directly on the meridian, as we can know from the preceding chapter. At this hour we have to take the altitude of a familiar fixed star and by it find the ascendant, just as chapter 3 showed us. The degree of the zodiac which falls on the meridian is the place of the moon or of the planet.

We can perform this operation for the moon during the day when it is at the meridian by using the sun in place of the fixed star. By night when the moon or the planet does not arrive at the meridian but appears toward the east or the west, we can work in the opposite direction: when we know that one of the fixed stars is at the meridian, we take the altitude of the moon or of the planet and make the figure according to the fixed star. Then we look at the part of the astrolabe where the moon or the planet is. The degree of the zodiac we find among the almucantars at the same altitude as the moon or the planet is the degree of the moon or the planet.

In this way we can find the place of the fixed stars which are not marked in the astrolabe by taking the figure of the sky, by whatever method, when the stars are at the meridian. We must understand this carefully, for the degree that we find thus is the one with which this star comes to the meridian. The declination of this star is easily known if we take note of the almucantars between the equinoctial and the altitude of the star when it is directly at noon.

15. *Pour savoir se les planetes sont septentrioneles ou meridioneles et aussi s'il sont retrogrades ou directes.*[96]

C'est assavoir que le soleil est tousjours endroit du milieu du zodiaque, c'est endroit de la ecliptique.[97] Mais les planetes sont aucune foiz devers septentrion du milieu du zodiaque et aucune fois devers midi. Et ceste variacion est appelee latitude[98] de planetes.

Quant il nous plaira ce assavoir nous devons savoir le lieu du planete par le chapitre ci-devant dit ou, plus certainement, par le almanach. Et devons actendre tant que le planete soit a midi ou bien pres, et doncques devons prendre sa hauteur. Et depuis devons tourner son degré au quel il est sur la lingne de midi. Doncques se la hauteur du planete est greigneur que la hauteur de son degré d'entre les almucantharat, le planete est septentrionel; et a tant que une hauteur est greigneur que l'autre, tant est le planete plus declinant en sa latitude. Mais se la hauteur du planete est mendre que la hauteur de son degré a midi, donques le planete est meridionel en sa latitude. Et ceste practique est molt a noter et a tenir a memoire.[99]

Et se il nous plaist a savoir se le planete est direct ou retrograde, nous devons considerer sa hauteur devers orient ou devers occident et aussi la hauteur de aucune estoile fixe qui soit pres du dit planete, et ces 2 hauteurs tenir en memoire. Et devons 4 nuitees ou 6 apres[100] considerer la hauteur du planete et de la dite estoile ainsi comme devant. Et se la hauteur du planete devers orient est mendre et devers occident greigneur que la hauteur de l'estoile, doncques le planete est direct et hastif; et se il avient au contraire, doncques le planete est retrograde. Et se les hautesses par pluseurs journees sont equales, doncques le planete est stacionere.[101]

16. *Quant[j] il nous convient faire la figure du ciel et les 12 maisons.*[102]

Ce chapitre est le plus hault et plus commun et souvent neccessaire sur toutes autres. Premierement pour ceste ouvrage devons trouver le ascendent en aucune maniere ainsi comme le 3ᵉ chapitre nous a monstré. Le quel degré ascendent doit estre mis sur l'orison orientel, et doncques le degré qui cherra sur la lingne de midi et celui qui cherra sur occident, le quel est tousjours opposit avecques le ascendent, et le degré qui cherra sur la minuit, qui est opposit[103] contre le midi, commencent[j] les quat[re] quartes et 4 maisons principaulz. C'est la premiere, la 10ᵉ, la 7ᵉ, et la 4ᵉ.[104]

Et doncques devons faire une figure[105] quarree ou reonde a nostre volenté, la quelle nous devons partir en 12 parties. Et escriptons le signe et le degré ascendent ou commencement de la figure devers la

15. *To know if the planets are northern or southern and also if they are retrograde or direct.*

We must know that the sun is always in the middle of the zodiac, that is, on the ecliptic. But the planets are sometimes to the north of the middle of the zodiac and sometimes to the south. This variation is called the latitude of the planets.

When we wish to determine this, we have to know the place of the planet, as described in the preceding chapter or, more certainly, by the almanac. We have to wait for the planet to be on the meridian or very near it before we take its altitude. Then we place its degree on the line of midday. If the altitude of the planet is greater than the altitude of its degree among the almucantars, the planet is northern; by as much as one altitude is greater than the other, by so much has the planet declined in its latitude. But if the altitude of the planet is less than the altitude of its meridional degree, the planet is southern in its latitude. This technique should be carefully noted and remembered.

If we wish to know if the planet is direct or retrograde, we have to consider its altitude toward the east or toward the west and also the altitude of some fixed star which is near this planet. We must keep these two altitudes in memory. Then four or six nights afterwards, we must measure the altitude of the planet and this star just as before. If the altitude of the planet towards the east is less and towards the west, greater than the altitude of the star, the planet is direct and rapid; if the opposite occurs, the planet is retrograde. If the altitudes are equal for several days, the planet is stationary.

16. *When we wish to make the figure of the sky and its 12 houses.*

This chapter is the most important and most often necessary of all the chapters. First, for this work we have to find the ascendant in some way, as chapter 3 showed us. This ascendant degree has to be put on the east horizon. Then the degree which falls on the line of midday, the degree which falls on the west horizon (which is always opposite to the ascendant), and the degree which falls on the line of midnight (which is opposite the line of midday), begin the 4 quadrants and the 4 principal houses, namely the 1st, the 10th, the 7th, and the 4th.

We must make a figure—square or round as we please—which we divide into 12 parts. We write the ascendant sign and degree at the beginning of the sign, towards the left hand. Its opposite on the

pote main; et son opposit a l'opposite partie en la figure, c'est la 7ᵉ maison; et le signe du midi en la 10ᵉ; et son opposit dedenz la 4ᵉ maison.

Ces 4 maisons principaulz ainsi ordenees les autres trouverons par telle maniere. Le degré ascendent doit estre mevé sur le bout de la 8ᵉ heure, et le signe et son degré qui cherra sur la lingne de minuit commence la 2ᵉ maison. Et apres devons mever le degré ascendent sur le bout de la 10ᵉ heure, et le lieu du zodiaque qui cherra sur la lingne de minuit commencera la 3ᵉ maison. Depuis devons mectre le opposit de l'ascendent, c'est le degré que nous avons devant mis sur la 7ᵉ maison, et doit estre mevé sur le bout de la 2ᵉ heure; et le lieu du zodiaque qui cherra sur la lingne de minuit commencera la 5ᵉ maison. Et se nous apres mevons le dit opposit degré de l'ascendent sur le bout de la 4ᵉ heure, ce qui cherra sur la lingne de minuit, commencera la 6ᵉ maisonᵏ. Ces maisons ainsi ordenees, devons faire le commencement de la 8ᵉ opposit a la 2ᵉ maison, et le commencement du neuvisme opposit a la tierce, et le commencement du 11ᵉ opposit a la 5ᵉ; et apres, le commencement du douzisme opposit au commencement du sisieme. Et ainsi avons fait et ferons les commencemens de 12 maisons sur chascun commencement naturel et a chascune heure.¹⁰⁶

17. *S'il nous plaist a savoir la hauteur du soleil par les heures equales ou inequales.*¹⁰⁷

C'est a savoir que aucune foizˡ est mestier ou esbatement savoir la hauteur du soleil ou d'aucune estoile comment que le soleil ne face point de clarté. Pour ce fault il savoir l'eure inequale ou equale par un horologe ou par la estimacion. Et se nous voulons ouvrer par les heures inequales, nous devons mectre le opposit du soleil sur l'arc de l'eure inequale ou le soleil meismes par nuit. Et en quel endroit de almucantharat cherra le lieu du soleil ou le lieu du planete ou de l'estoile, en tel endroit et hauteur sera le soleil par la hautesse, ou le planete ou l'estoile.

Mais s'il nous plaist a savoir la dicte euvre par les heures equales, que nous pourrons avoir par chascun commun horologe, nous devons mectre pour ce le degré du soleil sur la lingne de midi, se nous prenons le commencement des equales heures sur le midi, ou metons le degréᵐ du soleil sur la lingne de minuit, se nos heures equales sont commenciees de minuit. Et par quelconques maniere nous faisons, nous devons signer le almuri sur le lymbe du astralabe. Et de ylec devons tourner le rethe de astralabe selonc le mouvement journel, que le almuri se bouge pour chascune heure 15 degrés, tant de fois selonc ce que nous avons des heures equales. Et au bout du parfont de nos

opposite part of the figure is the 7th house; the sign of the meridian is in the 10th, and its opposite is in the 4th house.

These four principal houses being thus ordered, we find the others in the following manner. The ascendant degree must be placed on the end of the 8th hour line, and the sign and degree which falls on the line of midnight begins the 2nd house. Next we have to move the ascendant degree to the end of the 10th hour, and the point of the zodiac which falls on the line of midnight begins the 3rd house. Then we have to place the opposite of the ascendant (which is the degree that we earlier placed on the 7th house), and we move it to the end of the 2nd hour; and the place of the zodiac which falls on the line of midnight begins the 5th house. Afterwards, if we move this degree opposite the ascendant to the end of the 4th hour, what falls on the line of midnight begins the 6th house. These houses thus ordered, we have to take the beginning of the 8th to be opposite to the 2nd house, and the beginning of the ninth to be opposite the third, and the beginning of the 11th to be opposite the 5th, and afterwards, the beginning of the 12th to be opposite to the beginning of the 6th. Thus we make the beginnings of the 12 houses at each natural beginning and at each hour.

17. *If we want to know the height of the sun by equal and unequal hours.*

Sometimes it is interesting or necessary to know the altitude of the sun or some star although the sun may not be shining. In this case we must know the unequal or equal hour by a clock or by estimation. If we want to work with unequal hours we have to put the opposite of the sun, or by night the sun itself, on the arc of the unequal hour. Wherever the place of the sun (or planet or star) falls among the almucantars, that will be the altitude of the sun (or planet or star).

But if we want to work by equal hours, which we can obtain from any clock, we have to put the degree of the sun in this case on the line of midday, if we take the beginning of the equal hours to be noon; or we put the degree of the sun on the line of midnight if we begin the equal hours at midnight. By whatever technique we use, we must make a mark beside the almuri on the limb of the astrolabe. We then turn the rete of the astrolabe from this point according to the diurnal motion, so that the almuri moves 15 degrees for each hour, according to the total number of equal hours we have. At the end of our procedure

heures, le soleil et les lieus des planetes cherront en leurs parties du monde[108] et a leur hauteur d'entre les almucantharat ne plus ne mains comme se nous eussions ce trouvé par les estoiles meismes.

Et se nous voulons savoir a quans degrés le soleil ou aucune autre estoile se lieve ou baisse dedens une heure equale ou inequale, nous considerons le lieu du soleil ou son opposit selonc ce qui sera mestier; et le almuri nous signerons a nostre limbe et bougerons le rethe par 15 degrés pour une heure equale ou le degré opposit du soleil du commencement de l'eure inequale jusques a la fin. Et tant comme le soleil ou l'estoile se lieve ou avale d'entre les almucantharat, a tant se lieve ou baisse le soleil ou l'estoile a la dicte heure.

Ce sont les communs proffis et ouvrages du astralabe souvent cheans en besoingne et en practique de astronomie.

18. *Pour mesurer chascune chose haute en pluseurs manieres.*[109]

Pour la maniere de ce chapitre[110] devons savoir que les choses que nous voulons mesurer sont aucune foiz assisses en places plainnes telement que nous nous povons approchier et alongnier tant come nous voulons. Et aucune foiz sont eulz assises telement que nous ne n'en povons approuchier pour cause de yaue ou pour causes de fosses ou pour autres causes. Et pour ce est ce chapitre parti en 2 parties principales. Et est assavoir que toutes choses mesurees par ceste maniere doivent estre droitement drescees et non pendans en nul costé, et l'espace d'entre nous et les dictes choses doit estre plaine et non pendant. Avecques ce doit le mesureur estre acoustumé de ouvrer souvent et ce faire par bonne diligence.

Premiere partie

Se nous voulons mesurer aucune chose haute[111] a quoy nous pourrons approuchier et esloingnier tant comme il sera mestier, nous devons prandre le astralabe et nous dressier droitement et joindre nos piés ensemble et regarder parmi les pertuis, et donques de la riulle, diligenment la hauteur de la chose, <et nous esloigner ou approuchier devers la chose>[n] tant de foiz comme la riulle passe parmi le dyametre d'un des quarrés[112] de astralabe en regardant° la hauteur parmy lé pertuis. Et doncques devons noter la place du milieu de nos piés et avecques ce devons adjouster tant comme il a de nostre oeil a terre.[113] Et de ylec devons mesurer jusques au pié de la chose mesuree ou au lieu droitement subsistant le point plus haut de la chose. Et doncques la chose est droitement si haute comme il a du lieu ou nous

with the hours, the sun and the places of the planets will fall in their parts of the world and at their altitude among the almucantars just as if we had found it by the stars themselves.

If we want to know how many degrees the sun or any other star rises or descends in an equal or unequal hour, we consider the place of the sun or its opposite according to what is needed. We mark the almuri on our limb and move the rete by 15 degrees for an equal hour; or we move the degree opposite the sun from the beginning of the unequal hour to the end. As the sun or the star rises or falls among the almucantars, so the sun or star rises or falls at that hour.

These are the general uses and operations of the astrolabe often needed in the practice of astronomy.

18. *To measure the height of things by several methods.*

To understand this chapter we have to know that the things that we want to measure are sometimes located in flat places so that we can approach and move away from the objects as much as we want. Sometimes they are located in such a way that we cannot approach them because of a body of water, ditches, or other obstacles. For this reason, this chapter is divided into two principal parts. We must know that all things measured in this way must stand straight and must not lean in any direction and that the space between us and these things must be flat and level. And in addition, the measurer must be used to working persistently and very diligently.

First part

If we want to measure some high thing which we can approach and move away from as much as necessary, we have to hold the astrolabe and look through the holes while standing straight with our feet together. With the rule we must diligently sight the summit of the thing and move closer to or farther away from the thing until the rule sits upon the diameter of one of the squares of the astrolabe while we are sighting the height through the holes. We must take note of the place between our feet and add as much as there is from our eye to the ground. Then we must take the measurement to the foot of the thing measured or to the place directly under the highest point of the

preismes la hauteur avecques nostre hauteur[114] jusques au pié de la tour et de la chose mesuree.

Et apres devons mesurer par brassees ou par nos piés ou par aucune autre mesure a nostre volenté afin que nous puissons dire[115] combien la chose est haute. Et se nous povons monter en hault nous prouverons nostre art par une cordelete jusques a tant que nous soions bien certains et duis de nostre art et science.

C'est aussi asavoir que en regardant la hautesse parmy lé pertuis la riulle chiet sus un de costés ou sur l'autre de la quarreure,[116] les quelz sont chascun costé parti en 12 parties qui sont appelees pointes.[117] Et se la riulle chiet sur le costé de quarreuere prouchaine de nous, nous sommes trop pres de la tour. Et se ele chiet sur l'autre costé nous sommes trop loins.[118] Et pour ce a il encore une autre riulle, car se il nous plaist a mesurer sans ce que soit sur le dyametre[p] de la quarreure, nous devons veoir sur quel costé la riulle chiet et quantes pointes elle trenche et en quelle proporcion les pointes sont a 12: c'est a dire la mutet,[q] le tiers, ou le quart—comme 6, qui sont la muté de 12; et 4, le tiers; et 3, le quart.[119] Et en quelle maniere proporcion les pointes sont a 12 en telle proporcion se a la tour a l'espace d'entre nous et la tour en tres loing ou tres court.

En example, se la riulle chiet sur le costé du trop lonc et trenche 6 poins l'espace entor nous et la tour est doublé contre la hauteur, car 12 sont 2 fois tant comme 6. Mais se la riulle chiet sur le costé qui nous fait estre trop pres de la tour et trenche 6 poins, doncques la hauteur de la tour est 2 fois si haute comme l'espace est longue d'entre nous et la tour. Et tousjours devons adjouster la hauteur de nostre oeil de la terre.[120] Et ces 2 manieres de mesures et celles qui ainsi sont se font droitement par tele maniere par le quadrant et son quarré comme nous avons dist de astralabe.

Et nous devons bien garder[121] en cel ouvrage que le astralabe soit egalment pesant, de nul costé plus que de l'autre. Et ce pouvrons prouver par un plommet pendant parmi le astralabe. Et aussi devons estre certains que le quadrant et sa quarreure soit droitement partis, car autrement nous pourrons faillir pour cause[r] du instrument et non de nostre deffaute ou ouvrage.

Seconde partie [122]

Et se la chose est assise que nous voulons mesurer en tele maniere que nous ne pouvons approuchier a elle pour mesurer l'espace d'entre nous et le pié de la dite chose, doncques nous devons eslire une

thing measured. The distance from the foot of the tower or the thing measured to the spot from which we took its height (adding our own height) is the exact height of the thing.

Then we must measure in fathoms or feet, or some measure of our choice, so that we can say in numbers how high the thing is. And if we can climb to the top, we can test our art by means of a small rope until we are skilled and certain in our art and science.

We should also know that when we are taking a height through the holes, the rule falls on one side of the square or the other, of which each side is divided into 12 parts called points. If the rule falls on the side of the square near us, we are too close to the tower, and if it falls on the other side, we are too far away. There is therefore another procedure: for if we wish to measure without considering what falls on the diameter of the square, we must note on which side the rule falls and the number of points it crosses and in what proportion that number is to 12; that is, one half, one third, or one fourth (as 6 is half of 12, 4 is a third, and 3 is a fourth). In whatever proportion the points are to 12, the tower is in the same proportion to the space between us and the tower, whether longer or shorter.

For example, if the rule falls on the side of greater distance and crosses 6 points, the space between us and the tower is twice the height of the tower, for 12 is two times as much as 6. But if the rule falls on the side that we take to show greater nearness and crosses 6 points, then the height of the tower is two times greater than the space between us and the tower, and always we must add the distance from our eye to the ground. These two ways of measuring and those which are like them are done in just the same way on the quadrant and its square as we have said of the astrolabe.

We must make sure, in this operation, that the astrolabe is balanced, no side weighing more than the other. And this we can test by means of a plumb line hanging from the center of the astrolabe. Likewise we must be sure that the quadrant and its square are correctly divided, for otherwise we can make an error because of the instrument and not because of our technique.

Second part

If the thing we wish to measure is situated so that we cannot approach it to measure the distance between us and the foot of the thing, then we must choose an open, level place on which we can

place toute plainne et equale sur la quele nous nous puissons esloingnier et approuchier droitement devers la hautesse par aucune longueur de 10 ou 20 ou 30 piés loing ou environ. Et au bout de la place prochainne devers la hauteur devons regarder diligenment le point plus hault de la chose et devons aler devant ou derriere jusques a tant que la riulle trenchie aucune partie proporcionele de 12 poins,[123] c'est qu'il trenche ou 2 ou 3 ou 4 ou 6.

Et illec devons faire une note et merquer la place de nostre stacion premiere. Et devons tenir en memoire quantes fois les pointes prises sont dedens 12, et depuis devons droitement nous esloingnier de la hautesse et encore une fois prendre la hautesse parmi les pertuis si comme devant et bien tenir le nombre des poins et combien de fois elles sont en 12. Et a la seconde stacion faire aussi un signe. Et doncques devons oster le nombre combien de fois des pointes[124] premierement prises du nombre quantes fois de pointes secondement retenues. Et le demorant nous monstre en quelle proporcion l'espace d'entre nos 2 stacions sera a la hautesse. Car se il en demeure un, l'espace est equale a la hautesse. Et se il en demeure 2, l'espace est deux fois si longue comme la hautesse; et se il en demeure 3, elle sera trois fois si longue. Et adoncques devons mesurer l'espace d'entre nos 2 stacions par nos piés ou par aucune mesure a nostre volenté afin que nous sachons dire par mesure la hautesse mesuree sans ce que nous approuchiens a elle.

Example, j'ay mesuré une hautesse par ceste maniere. Premierement je pris les pointes parmy les pertuis de la hautesse, les quelz estoient 4 et sont trois fois en 12, si que j'en tenoie 3 en memoire, le quel nombre est appelé quatre fois[125] des premieres pointes. Et apres je me esloingnie de la hautesse par une droite lingne a tant que je regardai la hautesse parmy les pertuis, et la riulle trencha trois pointes, les queles sont quatre fois en 12, les quiex 4 je tenoie en memoire, et est le nombre combien de fois de la seconde stacion. Et doncques je ostay le nombre premier combien de fois qui estoit 3 du secont combien de qui estoit 4, et il m'en demoura 1, et pour ce l'espace entre mes 2 stacions estoit equale a la hautesse, la quele je trouvay par nombre de mesure tenir 20 piés, si que la hautesse mesuré tenoit 20 piés de hault comment que je ne puisse mie approuchier a elle. Et par semblabe example pourrons nous faire a chascune hautesse.

withdraw from and approach toward the thing by a distance of 10, 20, or 30 feet or thereabouts. Near the edge of the place closer to the thing, we must diligently observe the highest point of the thing; then we must move forward or backward until the rule crosses some proportional part of the 12 points, that is, until it crosses 2, 3, 4, or 6.

At this place we must note and mark the place of our first station and remember how many times the points taken go into 12. Then we must move directly away from the elevation, again take its height through the holes as before, remember the number of points and how many times it goes into 12, and also make a mark at the second station. Then we must subtract the number of how many times the points taken at the second station go into 12. The remainder shows us in what proportion the height will be to the space between our two stations. For if the remainder is 1, the space is equal to the height. If the remainder is 2, the space is twice as long as the height. If it is 3, it will be three times as long. Then we must measure the space between our two stations, by feet or some other measure of our choosing, so that we will be able to express the measure of the elevation without having approached it.

For example, I have measured an elevation by this method. First I took the points of height through the holes, which points were 4 and go into 12 three times, so that I held 3 in memory, which number is called four times of the first points. Then I withdrew from the elevation in a straight line until I saw the elevation through the holes and the rule cross three points, which go into 12 four times; I held 4 in memory, and that is the number of how many times the points taken at the second station go into 12. And then I subtracted the first number of how many times, which was 3, from the second, which was 4, and there remained 1; and therefore the space between my two stations was equal to the height, which I found by number to measure 20 feet, so that the elevation measured 20 feet in height, even though I could not in any way go up to it. And by this example, we can work with any elevation.

19. *Et quant nous voulons mesurer aucun prael ou jardin ou autre longueur de plainne terre.*[126]

Nous devons pour ce faire regarder le bout de la longueur de champ parmy les 2 pertuis et aviser quantes pointes trenche la riulle. Et en quelle proporcion 12 sont aus pointes, a tele proporcion est le lonc du champ a nostre hauteur de l'ueil de la terre. C'est a dire se les pointes prises sont 6 fois en 12, l'espace sera sis foiz si loing comme nostre hauteur est. Et par tele maniere ferons nous en quelconques proporcion de pointes a 12.

20. *Pour savoir la hauteur des choses par leur ombres qui sont de par le soleil par jour ou pour cause de la lune par nuit.*[127]

Nous devons actendre par jour jusques le soleil soit chait precisement 45 degrés et a celle heure prendre la longueur de l'ombre, car elle est ne plus ni mains si longue comme la chose dressee est, se ce est une yglise ou tour ou une montaingne ou quelconques autre chose. Et tant comme le soleil est plus bas de 45 degrés, les ombres sont plus longues que les choses qui les font. Et quant le soleil est plus hault de 45 degrés, les ombres sont plus courtes de leur choses. Et pour ce se nous prenons le point dedenz le quarré en astralabe ou en quadrant a chascune heure quant nous prenons la hauteur du soleil, nous pourrons savoir la hauteur des choses se nous comparason faison de dit pointes a 12 par tele maniere comme je ay dit a la premiere partie de 18e chapitre en parlant des costés de quarreure et leur 12 pointes en ce faisant bonne diligence.

Et ainsi ay je, Pelerin de Prusse, l'an 1362, le 9 jour de may, a l'eure de prime, par l'aide de dieu accompli les proffiz et chapitres de la practique de astralabe briefment et simplement tout seulement par usage de astralabe sans meller ouvrages et besoingnes estraunges par guise de calculacion, afin que les ouvrages soient simples et de chascune personne entendables.

Explicit

19. *When we wish to measure a field, garden, or other length of flat ground.*

For this procedure we must sight the end of the length of the field through the two holes and see how many points the rule crosses. Whatever proportion 12 is to the points, such is the proportion of the length of the field to the height of our eye from the ground. That is, if the points taken go into 12 six times, the space will be six times as long as our height. And we work in the same way with any proportion of points to 12.

20. *To know the height of things by their shadows cast by the sun in the day or by the moon at night.*

By day we must watch until the sun has reached exactly 45 degrees and at that time take the length of the shadow, for it is neither longer nor shorter than the thing is tall, whether it is a church, a tower, a mountain, or anything else. To the extent that the sun is lower than 45 degrees the shadows are longer than the things that cast them. When the sun is higher than 45 degrees, the shadows are shorter than the things. Therefore, if we note the point in the square of the astrolabe or in the quadrant at any time we take the sun's altitude, we can know the height of the thing if we compare these points to 12, in such manner as we have said in the first part of chapter 18 in speaking of the sides of the square and their 12 points.

Thus, I, Pèlerin de Prusse, at the first hour of May 9, 1362, with the help of God, finished briefly and clearly the procedures and chapters solely on the practical use of the astrolabe without mixing in recondite and arduous calculations, so that these operations remain simple and can be understood by everyone.

Explicit

Textual Notes

^a ouvrer. MS: *ouvrir.*
^b *rethe.* MS: *reche.*
^c *le.* MS: *les.*
^d The words in pointed brackets are inserted from the margin.
^e The words in pointed brackets are inserted from the margin.
^f *plus bas.* MS: *pluseurs.*
^g *3e chapitre.* MS: *3 chapitres.*
^h *declinacion.* MS: *declaracion.*
ⁱ *Quant.* MS: *Pour et quant.*
^j *commencent.* MS: *commencant.*
^k *6e maison.* MS: *6 maisons.*
^l *foiz* inserted from margin.
^m *ou metons le degré.* MS: *ou le metons le degré.*
ⁿ The words in pointed brackets are inserted from the margin.
^o *regardant.* MS: *regardent.*
^p *sans ce que soit sur le dyametre.* The manuscript seems garbled here. See explanatory note 119.
^q *mutet.* MS: *mutel.*
^r *cause* inserted from margin.

Explanatory Notes

The following notes, in addition to their explanatory function, are intended to facilitate comparison of Pèlerin's work with that of Messahalla, Chaucer, and others. For the latter purpose, several shortened forms of reference are used:

M = Messahalla, *Compositio et operatio astrolabii* as edited by Gunther in *Chaucer and Messahalla on the Astrolabe.*

M (Seld) = Messahalla, *Compositio et operatio astrolabii* as represented in Oxford, Bodleian, MS Selden Supra 78, fols. 51–70. See Appendix A, below.

Chaucer = *Treatise on the Astrolabe*, in the *Riverside Chaucer*, edited by Benson.

Llobet = treatises by Llobet de Barcelona and his circle printed in Millás Vallicrosa, *Assaig d'Historia de les Idees Fisiques i Matemàtiques.*

Raymond = the treatise on the astrolabe by Raymond of Marseilles edited by Poulle for *Studi medievali.*

Leopold = Leopold of Austria, *Li Compilacions de le science des estoilles*, edited by Carmody.

Thabit = *Astronomical Works of Thabit b. Qurra*, edited by Carmody.

Henry Bate = *Magistralis compositio astrolabij* Hanrici Bate, edited by Gunther in *Astrolabes of the World.*

---

[1] *practique.* Cf. M (Seld) II.1, rubric: "practica astrolabii."

[2] *pluseurs instrumens, aucunes en figure reonde.* The "round" (i.e., spherical) instruments would be celestial globes or armilliary spheres of one sort or another. Cf. the prologue to Oresme's *Traité de l'espere*, where he says that in order to make astronomy comprehensible, "les sages anciens composerunt, entre les autres, un instrument qui est appelé espere materiel ou artificiel, le quel on puet regarder tout entour, manier, & tourner, & consider en partie la descripcion & le mouvement du monde & du ciel." In St. John's College, Oxford MS. 164, fol. 1r, these words appear under a miniature portrait of Oresme's and Pèlerin's patron, Charles V, contemplating an armilliary sphere. (A fifteenth-century manuscript of the same work also contains a portrait featuring the same instrument, and that portrait is reproduced in Lutz, *Essays on Manuscripts and Rare Books*, 65.) Chaucer, too, refers to an instrument complemental or supplemental to the astrolabe that he calls a "spere solide" (*Astrolabe* I.17 and II.26). The treatise *De solida sphera* (early fourteenth century?) seems to refer to an instrument "solid" in the modern English sense (see Benjamin and Toomer, 19 n. 72, and North, *Universe*, 50 n. 15), but Robertus Anglicus defines *solidus* as "three-dimensional," so that, he says, it is not

superfluous to say, "Spera est corpus solidum" (Thorndike, *Sphere*, 145). North remarks, "I do not think that the question of whether [Chaucer] meant a globe or an armilliary sphere can ever be settled." The same could be said of Pèlerin, or perhaps he refers to both. Charles V kept "une espere materiel de cuyvre" in the upper (larger) study at the Hôtel Saint-Pol (Labarte, *Inventaire*, item 2268).

[3] *instrumens... en figure plate*. An astrolabe represents the circles of the celestial sphere ("cercles... imaginés en la figure de la neuviesme spere") in stereographic projection. Cf. M I, prohemium (195): "spera qui fuerit in extensa plano"; also M I.17 (211): "De projectione spere in planum." This sort of projection represents the celestial circles without distortion, as Pèlerin points out in the same paragraph, because the circles project as circles (and not, for example, as ellipses). On astrolabes and the theory of stereographic projection, see Neugebauer, *Isis* 40 (1949):241, and Thomson, ed., *Jordanus de Nemore*, 18–43.

[4] *Ptholomee*. Messahalla too cites Ptolemy as an authority on the astrolabe (Part I, prohemium, 195), although there is no mention of one in his famous *Almagest*. The instrument he mentions in his *Planisphaerium* is probably similar to but not identical with the one described by Pèlerin and Messahalla (North, "Astrolabe," 104). In medieval treatises on the astrolabe, Ptolemy is often cited as a principal authority (Millás Vallicrosa, *Assaig*, 1:274 and n. 98; Borst, *Astrolab*, 17 and n. 18). Cf. Hermannus Contractus, *De mensura astrolabii*, 404: "In metienda igitur subtilissimae inventoris Ptolemaei *Walzachora*, id est plana sphaera, quam *Astrolabium* vocitamus...."

In Pèlerin's milieu, as in others in the Middle Ages, Ptolemy was regarded as an Egyptian king learned in astronomy. Christine de Pizan so regards him and compares Charles V to him (*Livre des fais*, 1:45–46). According to the fifteenth-century writer Simon de Phares, Charles commissioned French translations of Ptolemy's astrological works *Tetrabiblos* (i.e., *Quadripartitum*) and *Centiloquium* (*Recueil*, 228). Ptolemy is sometimes depicted as wearing royal garb and holding an astrolabe, as in the example reproduced on the cover of *Les Tables alphonsines* (E. Poulle, ed.).

[5] *essenciels et accidenteles*. As applied to celestial circles, these terms distinguish the permanent circles from the variable. The ecliptic, defined by the sun's annual path, and the equator, defined by the equinoctial points and the poles and axis of the world, are called *essential*. The circles based on the horizon and meridian vary according to the observer's location and are called *accidental*. On an astrolabe the accidental circles ("almucantars," "azimuths") are engraved on plates ("tables") that can be removed and replaced when the observer moves to a different geographical latitude. See Pèlerin I.8–12. In a different but related use of the terms, the signs of the zodiac (sometimes called the "circle of signs") are said to have certain powers called essential because they inhere in the signs themselves, and certain other powers called accidental because they fall to the signs as a result of their position relative to the horizon and meridian of a given place at a given moment. The French version of Alchabitius's *Introductorium*, which follows Pèlerin's treatise on the astrolabe in St. John's manuscript, provides an example at fol. 39 v: "... nous avon la dit l'estre du cercle des signes essenciels, or dison apres le accidentel. Quer le cercle est figuré en toute heure en tele figure qui est devisee en 4 parties, les queles le cercle l'emispere [i.e., the horizon] et le cercle demi le ciel qui fait demi jour [i.e., the meridian]." The French text translates the Latin "circuli essentiali" and "accidentale" of John of Seville's Latin version of Alchabitius (fol. 29 v in Oxford, MS Selden Supra 78).

[6] *cercles... imaginés*. Cf. Oresme, *Traité de l'espere*, (fol. 6v): "Chascune cercle de l'espere est ymaginé comme un grelle line ou comme un tenue filet...." Oresme here is following Sacrobosco, *De spera* II (89): "Cum etiam omnis circulus in spera

... intelligitur sicut lines vel circumferentia...," so that "est ymaginé," being equivalent to "intelligitur," would imply knowledge grasped by the intellect rather than perceived directly by the senses. Of one celestial circle in particular, the meridian, Oresme writes, "Un autre cercle est en l'espere material & imaginé ou ciel ..." (*Traité de l'espere*, fol. 7 v): i.e., there *is* a circle in the physical instrument representing one that *is imagined* in the sky. Writing for an earlier, less sophisticated audience, Martianus Capella finds it proper to explain that the circles of the celestial sphere are incorporeal and that one should not be misled into thinking otherwise by the bronze instrument which represents them (*De nuptiis* Book 8.815 and 816, and see North, *Universe*, 22). Chaucer's account of the meridian is on this point quite clear: "Thys lyne meridional is but a description or lyne ymagined ..." (*Astrolabe* II, lines 1-2). Benjamin and Toomer render Campanus of Novara's celestial *ymaginationes* as "models" or "picturings of the arrangements of things" that are not proofs but are related to proofs as if they were conclusions drawn from them (*Theorica planetarum* I.98-100, and Commentary on I.96-101). R. W. Southern's more general definition of the function of *imaginatio* also fits the present case: "it had to establish the permanent features of recurrent sense impressions of the same object; and it had then to retain those features in a lasting image" (*Robert Grosseteste*, 40-41).

[7] *mon tresredoubté seigneur, tres haut et noble prince*. In the *Livret*, written in 1361, Pèlerin addresses Charles as "le tresexcellent & puissant prince & mon tresredoubté seigneur, mon seigneur Charles, ainsné filz du roy de France, duc de normandie & daulphin de Vennoys, du quel je estoie comme indigne et de ces mendres serviteurs pour le temps ..." (fol. 33 v). Pèlerin is writing, he says, "en la petit consergerie de l'Ostel de mon seigneur de Normandie, de costé Saint Pol lez Paris" (fol. 110v).

[8] *en langue françoise*, etc. Cf. Chaucer's apology for writing in plain English on a scientific and technical subject: "curious endityng and hard sentence is ful hevy at onys for such a child [as ten-year-old Lowys] to lerne" (*Astrolabe*, Prol., lines 45-46); also Jean de Meung's preface to his translation of Boethius, addressed to Phillip IV: "... Boece de Consolacion que j'ai translaté de latin en françois. Ja soit ce que tu entendes bien le latin, mais toutevois est de moult plus legiers a entendre le françois que le latin" (Dedeck-Héry, ed., *Mediaeval Studies* 14 [1952]: 168).

[9] *latin sur cé ordenes*. On the level of Latinity of Charles and his circle, see R. F. Green, *Poets and Princepleasers*, 149-53, and Laird, *Manuscripta* 34:172-73.

[10] *Ces profis ... ai je parti*. Cf. the opening words of *Practique* I, below: "les proffis devant dis." As a term for the rules or subsections of Part II, *profis* corresponds to Chaucer's word "conclusiouns" (*Astrolabe*, Prol. lines 69-71: "The secunde partie shal teachen the worken the verrey practik of the foreseide conclusiouns"). Messahalla has no such general term for subsections, but another Latin treatise on the astrolabe, by Llobet de Barcelona, uses the term *titulus*, as in "Iste titulus est quomodo debes in primis labore per astrolapsum ..." (290).

[11] *le instrument appelé "astralabe."* M explains (195) that *astrolabe* is Greek for "taking of stars" ("acceptio stellarum").

[12] *ront en sa figure plate*. See n. 3, above.

[13] *roe*. I.e., the rete. See I.15, below.

[14] *"visage" ou "face."* The side of the astrolabe to which Pèlerin refers is the one on which the "mother" ("mere") is located. (See I.15, below.) Hence Chaucer (I.15) calls it the "womb side." He does not, in the *Astrolabe*, call it the face or visage, but the *Equatorie*, which may be by Chaucer, does use both "visage" and "face" to designate one side of the instrument it describes.

[15] *"dos."* M (196): "dorsum"; Chaucer I.15: "bakside," II.2: "bakhalf"; *Equatorie*: "bakside." Cf. Campanus II.77: "in dorsa ... matris."

[16] *une hautesse petite*. M (Seld) b: "alhantabor, id est ansa"; Chaucer I.2: "a

maner toret [i.e., a sort of shackle or eye-bolt] fast to the moder of thyn Astrelabie."

[17] *un annel nommé "armille."* M (Seld) a: "armilla suspensoria ... et dicitur arabice alhantica [Gunther: alhahucia]"; Chaucer I.2: "ryng." This ring, attached to the "hautesse petit" mentioned above, is held on the "doit," as Pèlerin says, or on the "thombe of thi right hond," according to Chaucer, I.1.

[18] *"mere."* M (Seld) c: "mater"; Chaucer I.4: "moder." Cf. Llobet, 278: "... maior tabula que est quasi mater continet infra se tabulas climatus totius mundi." On the "tables of climates" or climate plates, see the following note.

[19] *tables.* M (Seld) d: "tabule"; Chaucer I.3: "thynne plates compowned for divers clymates." *Tabule* is the technical term for the removable discs or plates representing the "accidental" circles. See n. 6, above.

[20] *un cercle reont espés, nommé le "lymbe."* M. I.1 (195): "limbus." Chaucer in *Astrolabe* I.4 uses the term "bordure," but *Equatorie* uses "lymbe" in the same sense, namely "border, rim." This limb is a circular band affixed to the main disc of the astrolabe so as to form a raised border within which the smaller climate-discs fit. It is divided into 360°. Cf. Llobet, 278: "... glutinatum est *Alnogiza* [i.e., the limb] id est umbo qui altior levatus, circumdat estremitates ipsius tabule continens infra se ipsas tabulas, et ipsum *Alnogiza,* divisas est per CCCLX partes."

[21] *le "equinoctial."* M (Seld) d: "circulus equinoccialis"; Chaucer I., lines 14-15: "The myddel cercle in wydnesse, of these 3, is clepid the cercle equinoxiall," with the added explanation (I.17, lines 20-24) that "This same cercle is clepid also Equator, that is weyer of the day; for whan the sonne is in the hevedes of Aries and Libra, than ben the dayes and the nightes ylike of lengthe in all the world." Cf. Sacrobosco, *De sphera,* 86: "... appellatur equator diei et noctis, quia adequat diem artificialem nocti...."

[22] *"cercle de Capricorne ... du Cancre."* M (Seld) d: "circulus cancri ... capricorni"; Chaucer I.17: "cercle of Cancre ... Capricorne." Pèlerin's *Livret,* fol. 35v, gives the names of all twelve zodiacal signs "en françois," including "Escrivasse" for Cancer and "Chievre sauvage" for Capricorn. He uses the vernacular because the signs are named "selonc les ymages des estoiles fixes les quelles sont endroit du dit cercle." Cf. Chaucer I.21, lines 57-59: "the sterres that ben ther fixed ben disposed in signes of bestes or shape like bestes." Chaucer appears to be here following Sacrobosco's *De spera* (87), where it is said that the zodiac is called *zodiac* "propter dispositionem stellarum fixarum in illis partem ad modem huismodi animalium." Cf. also Robert Grosseteste, *De sphaera,* 15: "eius partes imaginibus sunt insignitae nominibus animalium nuncupatis." Oresme, by contrast (*Traité de l'espere,* fol. 6r), explicitly declines to translate the Latin names of the signs, on the ground that, as applied to sections of the zodiac, they mean absolutely nothing.

[23] *"lingne de midy," "lingne de minuit."* M (Seld): "linea meridiei," "medie noctis"; Chaucer I.4: "south lyne, or ellis the lyne meridional," "north lyne, or ellis lyne of midnyght." The *Equatorie,* referring to a comparable line on the equatorium, comments "which lyne is clepid in the tretis of the astrelabie the midnyght lyne. ..." D. J. Price, editor of the *Equatorie,* suggests (158) that if reference is here made to an English treatise on the astrolabe, Chaucer's must be the one intended.

[24] *l'autre trait travers lonc parmi la table,* etc. This east-west line is not described in the corresponding section of Messahalla—M (Seld) j—but cf. Chaucer I.5: "est lyne, or ellis the lyne orientale," "west lyne, or ellis the lyne occidentale." Cf. Llobet (277): "unam [linea] uocitant lineam equalitatis et ipsa est que currens *Almaserec Rikile Almagrib,* id est de oriente usque ad occidentem per medium punctam dividit tabulum per medium."

[25] *"almucantharat" en langue de Arabie.* M (Seld) e: "Almucantharath"; Chaucer I.18: "almycanteras." The word is uninflected in both Messahalla and Pèlerin. Almucantars are circles used in measuring altitudes, and Leopold (160) gives "cercle

de elevacion" as a synonym for "almucanarac." Thabit (141) explains the term: "Quod autem dicuntur almucantharath quasi semicirculi est propter astrolabium ubi quidam eorum non integri sed ut dimidii circuli depinguntur." (The Arabic word means "arc or arched bridge.") Messahalla, at I.12 (208) on the construction of almucantars, writes "almuthantharath, quos Latini vocant 'progressiones solis'."

[26] *"orizon oblique."* The horizon of any observer who is not on the equator is an "oblique horizon." (An observer on the equator has a "right horizon," meaning that the horizon line on his astrolabe would be a straight line.) An astrolabist at Paris would require a plate engraved for the oblique horizon there, while an astrolabist at Oxford would require a different plate. Compare:

I prove it thus by the latitude of Oxenford: understonde wel that the height of oure artik fro oure northe orisonte is 51 degrees and 50 minutes....
(Chaucer, *Astrolabe* II.22)

Item: je le monstre par example en l'orizon de Paris, ou le pole est eslevé aussi comme par xlix degrés. (Oresme, *Traité de l'espere*, fol. 8v)

Raymond (880) gives the latitude of Marseilles (45°). See also Pèlerin's I.14, below.

[27] *le cercle du ciel qui depart la muté visible de la invisible.* M (Seld) e says simply that the horizon "dividit enim 2 hemispheria." Chaucer I.18, lines 8–10, writes that the horizon is "the cercle that divideth the two emysperies," but adds, "that is, the partie of the hevene above the erth and the partie bynethe." Cf. Sacrobosco, *De spera*, 91: "Orizon vero est circulus dividens inferius emisperium a superiori...." Pèlerin is actually a bit closer to Grosseteste, *De sphaera*, 16: "Horizon vero est circulus, qui dividit medietatem coeli visam a medietatem non visa...." Cf. also Oresme, *Traité de l'espere*, vol. 7v: "Encor ha un autre cercle en l'espere qui est appellé orizon, le quel devise la partie du ciel que nous voions sus nous d'aveque celle que nous ne voions pas et qui est aussi comme sous nous."

[28] *"orientale," "occidentale."* This explanation has no parallel in Messahalla, and Chaucer's explanation of the same point is somewhat different. Pèlerin says that the side of the astrolabe "devers la pote main" is called "orientale" and that the other side is called "occidentale." Chaucer I.6 says, "The est syde of thyn astrolabie is clepid the right syde, and the west syde is clepid the left syde. Forget not thys, litel Lowys. Put the ryng of thyn Astrolabie upon the thombe of thi right hond, and than wol his right side be toward thi lift side, and his left side wol be toward thy right side." Thus in Chaucer's terms the right side *of the astrolabe* is called the east side, whereas in Pèlerin's, the left side *of the astrolabist* is the east.

[29] *"cenith."* M (Seld) f: "cenit"; Chaucer I.18: "cenyth." The word here means "zenith" in the modern English sense. Elsewhere Messahalla and Chaucer use the word in a different sense. See n. 80 on *Practique* II.10, rubric.

[30] *instrumens ... portatis*. Pèlerin here departs from Messahalla to add an account of a portable astrolabe that, because of its smaller size, represents almucantars not for every degree but for every other degree or every third, fourth, or fifth degree. Chaucer, who is likewise writing about an "instrument portatif" (*Astrolabe*, Prol., line 73), also inserts at I.18 an appropriate account of the almucantars shown on such an instrument: "These almycanteras ben compowned by 2 and 2, all be it so that on diverse Astrelabies somme almykanteras ben divided by oon, and some by two, and somme by thre, after the quantite of the Astrelabie." Raymond, 882, likewise allows for less complete markings according to the size of the instrument.

[31] *"azimuth."* M (Seld) g: "azimuth"; Chaucer I.19: "azimutz." Azimuths (Arabic: al-sumūt, "directions") are great circles passing through the zenith and nadir. But on the astrolabe, as Pèlerin points out, they extend only from the zenith to the horizon. Cf. Leopold, 60: "Azimus sont 360 cercles qui issent hors dou pol de l'orizont par mi orizont, et cil sont appiellé cercle de direction."

³² *Et doivent estre a un parfait astralabe.* Here again, as in I.11, above, Pèlerin, departs from Messahalla to insert a remark distinguishing between a large "parfait astralabe" (marked with 360 azimuth lines, or 90 to the quadrant) and the smaller instrument with which Pèlerin is chiefly concerned. Cf. Chaucer I.19, describing a small instrument marked with one azimuth line for every 15 degrees: "And these same strikes or divisiouns ben clepied azimutz, and thei dividen the orizounte of thin Astrelabie [that is, all 360° of it] in 24 divisiouns." The plate shown in App. B, fig. 1, has 36 divisions.

³³ *12 traiz avecques la lingne de minuit.* I.e., twelve arcs plus the line of midnight, as shown in App. B, fig. 1.

³⁴ *heures inequales.* M (Seld) h: "hore"; Chaucer I.20: "houres of planets," although Chaucer elsewhere (II.8, II.10) uses the phrase "houres inequales" to mean the same thing. The term "hore" is ambiguous, and so Pèlerin specifies what kind of hours by using the phrase "heures inequales" to name a concept he explains at the beginning of II.3, below. Chaucer resolves the ambiguity of "hore" by using the phrase "houres of planets," and in so doing he places the concept in an astrological context not invoked by Pèlerin. Chaucer explains that context in *Astrolabe* II.12, "Special declaracioun of the houres of planets."

³⁵ The corresponding sections of Messahalla and Chaucer contain no explanations equivalent to the present one in Pèlerin as to why astrolabes are equipped with several plates marked for different geographical latitudes and how to choose the appropriate one. The information in Pèlerin's explanation is implicit, however, in a later section of Messahalla—namely II.23 ("De noticia tabule almucanterath")—which Chaucer translates as his II.21 ("To knowe for what latitude in eny regioun the almykanteras of eny table ben compowned"). Also, see *Practique* I.8 and n. 26, above.

³⁶ *l'estoile marine.* The North Star, i.e., the star alpha of the constellation Ursa Minor, called "stella maris" because of its usefulness in navigation (Allen, *Star-Names and their Meanings,* 454). Neither Messahalla nor Chaucer uses the term, but Raymond does (897: "maria stella"). Chaucer, II.22-24, uses the "pool star" for determining the observer's latitude.

³⁷ *"reys" ou "rethe."* A star-map represented on a metal plate perforated so that it vaguely resembles a net. The word comes from the Latin *rete,* meaning "net." In the Latin version of Messahalla *rete* is used in several places, but in the section Pèlerin is following—M (Seld) k—it is called by the Arabic name for "spider": "alhancabuz [for *ankabút*], id est aranea." Chaucer, I.21, lines 1-2, calls the object a "riet" and says that it is shaped like a net or like a spider's *web.* Chaucer's account is considerably longer than Pèlerin's, for he has supplemented Messahalla's description of the rete with Sacrobosco's description (87-89) of the celestial zodiac, which is represented on the rete. Cf. Llobet (278): "*Alhanicabut,* hoc est rete perforatum in quo sculpta sunt nomina casarum et stellarum fixarum." A rete is shown in App. B, fig. 2.

³⁸ *la partie du dit reys enclos d'un cercle.* This circle, traced by revolution of the heads of Aries and Libra, is the equinoctial circle, i.e., the celestial equator. Like most retes, the one shown in App. B, fig. 2, does not fully represent this circle, because to do so would obscure too much of the climate plate, which lies beneath the rete.

³⁹ *"septentrionel," "meridionel."* On the limit separating northern from southern stars Messahalla is a little confusing, for he mentions, at M (Seld) k, both the equinoctial ("motum captis arietis et libra") and the zodiac, together with the ecliptic ("via . . . solis"). Pèlerin therefore says there are two manners in which the terms "northern" and "southern" are used. In one manner they refer to stars north and south of the equinoctial ("un cercle qui se fait en tournant le reys du commencement du chief de Aries et de la Libre"). In another manner they refer to stars

north and south of the ecliptic ("circumference de zodiaque," i.e., "ecliptique"). Chaucer, I.21, lines 10–23, describes only one manner: stars within the zodiac of the astrolabe (and hence north of the ecliptic) are "sterres of the north," and those outside are "sterres of the south." He must then explain that some stars of the south are north of the equinoctial: "But I seie not that thei [i.e., stars of the south] arisen alle by southe of the est lyne; witnesse on Aldeberan and Algomeysa."

[40] *Alhabor, Altaire, Aldeberan.* Fixed stars often marked on the retes of astrolabes because they are "grans et notables." In describing the rete, M (Seld) k names no particular stars marked on it. Chaucer's corresponding description, at I.21, mentions Aldeberan (alpha Tauri), and in another context (II.3, line 47) he mentions Alhabor (Sirius, the Dog Star). Altaire (alpha Aquilae) is mentioned by Leopold, I.2, line 96. All three stars are marked by star-pointers ("longuetes") on the late-fourteenth-century French astrolabe pictured in Plate LXXIX of Gunther's *Astrolabes of the World* 2:342.

[41] *au commencement de Capricorne.* Both M (Seld) l and Gunther, 217, read "circulum capricorni," but Pèlerin and Chaucer (I.21, line 92: "heved of Capricorne") write as if their texts of Messahalla read "capitem capricorni." Their statements are accurate, and perhaps each independently improved upon Messahalla. Cf. Llobet, 279: "*Almeri,* hoc est eminens . . . supra caput capricornii."

[42] *"almuri," c'est a dire le monstreur.* M (Seld) l: "almuri, quod ostensor dicitur latine, denticulus scilicet"; Chaucer I.21, lines 90–1: "Thin almury is clepid the denticle of Capricorne, or ellis the calculer." *Almuri* is the Arabic *al-murī,* "the indicator, the little hand." Our reading of Messahalla differs slightly from Gunther's here, in that we construe the phrase "in alhantabuz relictus" as part of our section l (as suggested by the manuscript's capitalization of "Deinde" at the beginning of M [Seld] m) rather than as part of the following sentence. Hence section l should be read as saying that the almuri is a bit of metal left to project from the rete at the beginning of Capricorn. As W. W. Skeat explains (*Complete Works of Geoffrey Chaucer* 3;187), "the edge of the rete is cut away near the head of Capricorn, leaving only a small pointed projecting tongue which is called the almury or denticle." Cf. Llobet as quoted in the preceding note. With Chaucer's use of the term *calculer,* cf. Raymond, 885: "Almuri arabice, id est calculator." Raymond's editor, Poulle, notes that this use of *calculator* is rather rare in Latin and may be due to Raymond.

[43] *cheville . . . "cheval."* M (Seld) m: "equus" "cuneus;" cf. Gunther's edition, 217: "alphaeraz, id est equus"; Chaucer I.14: "hors" "wegge." In another place (417) Messahalla says the wedge may be made in various shapes, whereas Pèlerin here says it may be called by various names. Cf. Raymond, 889, who says the wedge may have a horse's or a cock's head: "clavux latus equum vel gallinatum habens caput." Llobet, 279, simply says "*Alfarat,* id est clavellus." This pin actually fits like a cotter-key into a hole or slit in the axial pin that holds the whole assembly of rete, plate, and mother together.

[44] *2 manieres de cercles.* M (Seld) m: "2 circuli equacionis solis." Chaucer distributes his discussion of these circles into four "conclusiouns" I.7, 8, 9, and 10. There are two concentric circles on the back of the astrolabe Pèlerin is describing. The inner one shows the months and days of the year, and the outer shows the signs and degrees. These circles are useful in converting the sun's position into time, the inner circle showing the date from the sun's annual motion and the outer showing time from its diurnal motion. (Cf. Chaucer I.7: ". . . every degree of the bordure conteneth 4 minutes; that is to seien mynutes of an houre.") Such conversion is probably what Messahalla means by his phrase (not translated by Pèlerin or Chaucer) "equation of the sun," which should not, therefore, be confused with *equatio solis* as used by John of Saxony, for example, to refer to the procedure for converting the sun's mean *motus* to true *motus* (*Les Tables Alphonsines* 17, line 17, p. 66).

Pèlerin adds from M (Seld) o information on a different function: the outer circle, in addition to being divided into signs of 30° each, is divided "par un autre nombre" from 0 to 90 in each quadrant for the taking of altitudes.

[45] *un demy quarré.* M (Seld) p: "quadrans"; Chaucer I.12: "skale in manere of 2 squyres." The "demy quarré" is a rectangle half as high as it is wide. See App. B, fig. 3. *The Riverside Chaucer,* 664, fig. 1, reproduces an Oxford manuscript illustration in which, following an error in Chaucer's text (I.12, lines 7-9), the labels "vmbra recta" and "vmbra versa" are switched. A large color photograph of the same page (MS Rawlinson D. 913, fol. 24r) is printed in *Aramco World Magazine* (May-June, 1982): 31.

[46] *cercles qui s'assemblent parmi le trou de l'instrument.* Pèlerin's addition to his source. See App. B, fig. 3.

[47] *la reigle:* M (Seld) q: "regula"; Chaucer I.13: "reule." See , App. B, fig. 4. Elsewhere both Pèlerin and Messahalla call the rule on the back of the astrolabe an *alidade*—Pèlerin at II.8 ("la alidade") and Messahalla at II.11 ("alidadam," "alidada"). Cf. Llobet (279): "... *Ahyldada* hoc quo iacet supra dorsum astrolapsus cum quo accipient altitudinem, et habent duo capita et recta per foratum per que foramine deprehendunt gradus solis, et probant stellas...." Chaucer (I. 22) and the *Equatorie* (consistently throughout) call it the "label." It is used both for sighting, through the holes in the vanes, and for pointing to degrees, months, and days in the border. (See *Practique* I.18 and n. 44, above.)

[48] Cf. M II.1; Chaucer II.1.

[49] *sus le dos de l'instrument:* Chaucer II.1, line 9: "in the bakhalf of myn Astrelabie." This specifying of which side of the astrolabe to use does not occur in the text of Messahalla but does occur in a marginal note in M (Seld), a fact which has encouraged the suggestion that Chaucer read Messahalla in that manuscript (Masi, *Manuscripta* 19:41-42).

[50] *lieu du soleil:* M II.1: "graduum solis"; Chaucer II.1, lines 4-5: "the degre of thy sonne." Here, as elsewhere in Pèlerin's treatise, "lieu du soleil" refers to the sun's ecliptic longitude, i.e., its location in the zodiac.

[51] *almanach:* Cf. Chaucer, Prol., line 93: "almenak." Both writers refer to an ephemeris, a set of tables giving planetary positions for a given year or period of years, usually at five- or ten-day intervals.

[52] *a semblabe maniere:* Just as we can find the date if we know the sun's longitude, so *in a similar way* we can find the sun's longitude if we know the date.

[53] Cf. M II.2; Chaucer II.2.

[54] *un de noz doiz.* Cf. M II.2: "de manu tua dextra"; Chaucer II.2: "upon thy right thombe."

[55] *lever et baissier,* etc. Our text here is a reconstruction of some imperfect lines in the manuscript. The words we have enclosed in pointed brackets are written at the very top of the manuscript page and marked for insertion at the point where we have placed them.

Given the nature of the instrument and the problem to be solved, it is clear that one must adjust the rule (alidade) so that a ray of the sun shines through the holes in the vanes affixed to each end of the rule, as directed by the words of Messahalla which Pèlerin is translating: "... subleva vel deprime regulam, donec radius solis per utriusque tabule foramen transeat" (II.2).

[56] *cours.* I.e., location. As an astronomical term, the Latin word *motus* usually means not "motion" but "angular position." If a translator ignores the technical meaning, he may render *motus* as *cours,* as does the translator of Leopold's *Compilacion.* (See Carmody's introduction to the *Compilacion,* 45.) Pèlerin is not translating *motus* here but is rendering the idea of it by the term *cours.*

⁵⁷ *les 4 tros.* As explained below, the vanes on some astrolabes are pierced by two sets of holes, a smaller pair and a larger pair.

⁵⁸ *2 paires de tros.* Cf. M I.4 (199) on the construction of the alidade: "... et in unaquaque tabula sint duo foramina, maius scilicet et minus, minus ad accipiendum radios solis in die et maius ad accipiendum stellas in nocte."

⁵⁹ *2 petis dentelés.* Some alidades, though not the one in fig. 4 of App. B, have vanes equipped with projecting points or notches, like front and rear gun sights.

⁶⁰ Cf. M II.3; Chaucer II.3. Pèlerin's II.3 incorporates the basic procedures set forth in his II.1 and II.2 and solves two fundamental astrolabic problems: finding the time and finding the ascendant (i.e., the degree of the ecliptic on the eastern horizon). It is therefore worth explaining in some detail. Following the directions in II.1, one finds the sun's ecliptic longitude (*lieu du soleil* or *signe et degré du soleil*) either by reading it off the back of the astrolabe (*dos*) or by using an ephemeris (*almanach*). Then, still using the back of the astrolabe, one finds the sun's angular altitude (*hauteur*) above the horizon by holding the astrolabe by the suspending ring (*armille*), lining up the moveable alidade (*riulle*) with the sun, and reading the degrees off the graduated rim (*lymbe* or *bort*). On the other side of the astrolabe (*visage* or *face*), one rotates the rete (*reys, rethe*) until the longitudinal degree of the sun lies over the almucantar representing the sun's altitude. When the rule is placed over the sun's position, it points to the time on the rim, with the noon at the top (*lingne de midi*), midnight at the bottom (*lingne de minuit*), 6 A.M. to one's left (*lingne de vray orient*) and 6 P.M. to one's right (*lingne de vray occident*). The degree of the ecliptic (on the rete) which touches the east horizon (*orizon occidentel*) establishes the ascendant and hence the first astrological house (*maison*), from which the remaining houses can be computed, as explained in II.16, for the casting of a horoscope.

Pèlerin prefaces all this with a definition of equal and unequal hours which Chaucer, following Messahalla's plan more closely, postpones until II.10, 11 (M II.6, 8).

⁶¹ *le soleil et son nadair, c'est a dire son opposit.* M II.3: "nadays," but "nadair" in M (Seld) II.3. The Arabic is *naẓīr* (Kunitzsch, "Glossar," 542). In reading the hour from the sun's nadir, Pèlerin remains faithful to Messahalla's procedure, whereas Chaucer departs from it and arrives at the same result by counting the hours from the line of midnight to the sun's position.

⁶² *nulle heure du jour est tant a doubter,* etc. This warning against basing the procedure on an altitude taken near the meridional line is not present in any version of Messahalla we have seen, but it has an equivalent in Chaucer, II.3, lines 63–81:

> But natheles this rule in generall wol I warne the for evere: Ne make the nevere bold to have take a just ascendent by thin Astrelabie, or elles to have set justly a clokke, whan eny celestial body by which that thou wenyst governe thilke thinges be nigh the south lyne. For trust wel, whan the sonne is nygh the meridional lyne, the degre of the sonne renneth so longe consentrik upon the almykanteras that sothly thou shalt erre fro the just ascendent. The same conclusioun sey I by the centre of eny sterre fix by niyght. And more over, by experience I wot wel that in oure orisounte, from xi of the clokke unto oon of the clokke, in taking of a just ascendent in a portatif Astrelabie it is to hard to knowe—I mene from xi of the clokke before the houre of noon til oon of the clokke next folewyng.

Cf. Henry Bate: "... cum stelle per quam volumus negociari est prope circulum merediei ... tunc eadem elevatione diu durat."

⁶³ Cf. M II.4; Chaucer II.6.

⁶⁴ *le vespertin crepuscle.* Pèlerin omits the morning crepuscule. Cf. M II.4: "finem crepusculi vespertini et inicium matutini"; Chaucer II.6: "the spryng of the dawenyng and the ende of the evenyng."

⁶⁵ *nadir du soleil.* Since almucantars are not marked on the astrolabe below the horizon (see I.5, above), the sun's nadir is set on the proper almucantar above the horizon, thus causing the sun's position to be properly located below the horizon.

⁶⁶ *18 almucantharat.* Both Pèlerin and Chaucer follow Messahalla in setting the crepuscular limit when the sun is 18° below the horizon, although another of Chaucer's authorities, Nicholas of Lynn (cited in Chaucer's *Astrolabe* Prol., line 86) sets it at 20° (*The Kalendarium of Nicholas of Lynn,* ed. Eisner, 12).

⁶⁷ Cf. M II.5, 13; Chaucer II.7, 15.

⁶⁸ *l'eure... du jour et de la nuit artificiele.* This is the "heure inequale" defined at the beginning of II.3. The term *artificial day* (Chaucer II.7: "day artificiall"), for the period from sunrise to sunset, was standard from the early thirteenth century, as in Grosseteste, *De sphaera,* 22, and Sacrobosco, *De spere,* 102. A *natural day,* by contrast, is the period of one complete revolution of the firmament—which period, divided by 24, yields *equal hours.* With Pèlerin's formulations, cf. Oresme, *Traité de l'espere,* fol. 10v: "Par le jour artificiel l'en entent le temps du solail levant jusques a soleil couchant": and (fol. 9v): "Le jour naturel est l'espace de temps en la quelle le soleil fait 1 tour entier entour la terre." Also, cf. the English translation of William of England's *De Urina non visa:* "natural day contaynynge in hit silf 24 houres, that is a day artificial and a nyght" (Cambridge, Trinity College, MS. 0.5.26, fol. 31r).

⁶⁹ *noter le almuri au droit du limbe.* M II.5: "nota locum almuri inter gradus limbi." Chaucer's method is slightly different in that he uses the rule ("label") as a pointer instead of the almuri: "... ley thy label on the degree of the sonne, and at the point of thy label in the bordure set a pricke" (II.7, lines 20-24).

⁷⁰ The matter in this paragraph is inserted from M II.13, which Chaucer, II.15, renders somewhat more clearly: "Loke which degrees ben ylike fer from the hevedes of Cancer and Capricorne, and loke when the sonne is in eny of thilke degrees; than ben the dayes ylike of lengthe. This is to seyn that as long is the day in that month, as was such a day in such a month...." Cf. also M II.15 and Chaucer II.16.

⁷¹ *Et en semblabe maniere,* etc. The reference is to the manner of working described in the first paragraph of this section.

⁷² Cf. M II.8; Chaucer II.11.

⁷³ *tant de fois que le almuri passe 15 degrés du limbe.* M II.8 presents the problem as one of arithmetical division: "... diuide gradus qui sunt inter .2. notas per .15. et habebis horas equales." Pèlerin uses mechanical means so that counting is all that is required. Chaucer, II.11, notes that since the limb or border of his astrolabe is already marked to represent hours (15° to an hour), the solution is intuitively clear: "The quantite of houres equales, that is to seyn houres of the clokke, ben departed by 15 degrees alredy in the bordure of thin Astrelaby, as wel by night as by day, generaly for evere. What nedith more declaracioun?"

⁷⁴ Cf. M II.6, 7; Chaucer II.10. Messahalla's method for finding the length of unequal hours by day is to divide the arc of day by 12 arithmetically: "... diuide arcum diei per .12., et habebis numerum graduum hore diurne...." Chaucer II.10 follows Messahalla closely, but Pèlerin solves the problem by an entirely different method, one which, like that in II.6, avoids arithmetic. The result of both methods is to express unequal hours in equinoctial degrees, in which equal hours are measured at 15° each. Llobet (285) describes the method Pèlerin adopted, except that Llobet is converting unequal hours, instead of one unequal hour, into equinoctial degrees. He calls unequal hours "horas tortas" and equal hours "horas rectas":

> Quando queris tornare horas tortas ad horas rectas per astrolapsum, accipe quot queris, et in ultima linea horarum quas acceptisti, pone nadair solis, et uide ubi stat almeri, et pone ibi signum, postea circum uolue ipsum nadair solis ab ultima linea usque ad primum almucantarat prime hore et uide ubi

stat almeri, et pone ibi signum et ipsos ordines quos ambulabat almeri, partire per ordinem rectarum horarum, id est per xv, et uidebis quot inde colligis horas rectas.

[75] Cf. M II.10; Chaucer II.13; also M II.12 (omitted by Chaucer). Messahalla's calculations are based on the position of the sun alone, although he adds at the end of II.10, "Similiter fac cum stellis fixis." (Cf. Chaucer II.13: "So may thou know in the same line the heighest cours that eny sterre fix clymbeth by night.") Pèlerin creates mild confusion by using the longitude of the sun, or any other longitude, or any fixed star marked on the astrolabe.

[76] *hauteur de midi.* I.e., the sun's meridional altitude. Cf. M II.12: "altitudinem solis meridianam."

[77] *le entier compas des heures sur le dos figurees.* I.e., the sixth hour line, representing the meridional circle. Of the hour lines engraved on the back of the astrolabe, only the sixth is represented as a complete circle, the others being partial circles. See App. B, fig. 3. Not all astrolabes are marked in this fashion (see North, "Astrolabe," 105, showing no complete circle), but evidently Pèlerin's was.

[78] *le quadrant.* The instructions in this paragraph are for using the astrolabe in the same way one would use the instrument called a "quadrant" (M II.12: "quadrante") to find the time in unequal hours. On the development of the quadrant out of attempts to increase the observational accuracy of the astrolabe, see Poulle, *Journal des savants* (avril-juin, 1964): 148-49.

[79] Cf. M II.16; Chaucer II.18.

[80] Cf. M II.17; Chaucer II.33, and M II.18; Chaucer II.31. The phrase *en quel endroit* corresponds to M II.17: "cenith"; Chaucer II.33: "cenith," i.e., azimuth. The azimuths mark directions north, south, east, and west, and points in between. Chaucer prefaces his rendering of M II.15 (Chaucer II.31) with the relevant explanation that "the sonne ariseth not alway verray est, but somtyme by northe the est and somtyme by south the est."

[81] *comme nous faisons pour trouver les heures et le ascendent.* See II.3, above.

[82] Cf. M II.29; Chaucer II.28.

[83] *signer le lieu.* Chaucer II.28, line 4: "set ther a prikke." North, *Universe,* 68, n. 25, remarks that "Messahalla does not go into such small details as pricks of ink; but cf. Raymond, quoted in the following note: "nota signum in limbo" and "signum facies."

[84] Pèlerin, like Chaucer II.28, lines 1-11, gives step-by-step instructions, similar to those in Raymond, 894:

... primum pone uniuscujusque gradum signi super almucantarath primum [i.e., "l'orizon"] et nota signum in limbo ubi fuit almuri in principio arcus; ducens ergo rete usque quo ejusdem signi gradus ultimus ad almucantarat predictum deveniat, videbis postmodum ubi fuerit almuri secundo et signum facies; quot enim gradus a priori signo usque ad posterius almuri pertranseat, tot gradibus in ea regione signum illum ascendit. Sic de ceteris facere momento.

Messahalla, by contrast, simply says "moue rethe ab inicio signi usque ad finem eiusdem, et gradus quibus mouetur in margine almuri erunt ascensiones signorum."

[85] *oblique ou de courtes ascencions.* Chaucer II.28 identifies these as the signs from Capricorn through Gemini and calls them "tortuous" or "croked."

[86] *droites et de longues ascencions.* Chaucer II.28 identifies these as the signs from Cancer through Sagittarius and calls them "right."

[87] The matter in this paragraph does not appear in the corresponding section of Messahalla, but cf. Chaucer II.28, lines 22-25.

[88] Cf. M II.30. This chapter of Messahalla is not represented by Chaucer, but an

English paraphrase of it is added to Chaucer's II.3 in MS Bodley 619. It is printed by Gunther, *Chaucer and Messahalla*, 182 n. 1.

[89] *sur les almucantharat en la partie ou elle est.* I.e., on the almucantar representing the star's altitude and on the appropriate side of the meridian. Cf. M II.30: "super similem altitudinem."

[90] Cf. M II.19; Chaucer II.29. Actually, only the instructions in the first paragraph of II.13 correspond to M II.19 ("De quatuor plagis mundi") and to Chaucer II.29 ("To knowe justly the 4 quarters of the world, as Est, West, North, and South").

[91] *contre le midi devers minuit.* Since the midnight line is oriented to lie northward from the "bastonnet," the shadow of the noon sun falls on that line and hence opposite the meridional line, which lies southward from the bastonnet.

[92] Cf. M II.31, 32; Chaucer II.34. Despite the rubric's claim to deal only with finding the place of the moon or other planets, Pèlerin actually deals both with this problem and with finding the place of fixed stars not marked on the rete of the astrolabe by "pointers" ("longuetes": see I.15). He is combining M II.31 ("De noticia stellarum incognitarum non positarum in astrolabio") with elements from M II.32 ("De loco lunae vel cuiusque planete"). The principle is that when the moon or other planet or a fixed star not on the rete is on the meridian, then whatever degree of the ecliptic is also on the meridian marks the position of the body in question. The problem is to set the rete. This is done, Pèlerin says, by taking the altitude of some body locatable on the rete—a fixed star that *is* marked by a pointer, or the sun at a known longitude—and setting the rete accordingly. For example, one takes the altitude of a fixed star marked on the rete and sets the rete so that the star lies on the almucantar representing that altitude (on the appropriate side of the meridian). Then the degree of the ecliptic which lies on the meridian marks the position of that body on the meridian whose position we wish to know. There is a caveat to all the above: what we have been calling position ("lieu," "en quel signe il sont et en quel degré") is not properly ecliptic longitude unless the body in question has no latitude from the ecliptic. It is rather what used to be called *mediation* (Saunders, *All the Astrolabes*, 5–6), that is, the degree of the ecliptic that comes to the meridian at the same time as the moon or other planet or fixed star. (See the first paragraph of Pèlerin's II.15, below.) This caveat is what Pèlerin offers at the end of II.14, and that same consideration prompts Chaucer to warn (II.34, lines 13–16) that "Comoun tretes of the Astrelabie ne maken non excepcioun whether the mone have latitude [from the ecliptic] or noon...." Masi points out that a similar warning is written in the margin of M (Seld) beside Messahalla's II.34 (*Manuscripta* 19:42).

[93] *la lune quant elle est a midi.* Cf. M II.32: "Si autem apparuit in die...," and Chaucer II.34: "And *nota* that yf the mone shew himself by light of day...."

[94] *ferons la figure.* I.e., set the rete.

[95] *declinacion ... est ... legierement a cognoistre.* Cf. M II.31: "latitude patet." We have made an emendation here, for where the manuscript clearly reads *declaracion*, the context just as clearly requires *declinacion*, as equivalent to Messahalla's "latitudo." *Latitude* and *declination* were still ambiguous, both being used at times to refer to angular distance from the ecliptic and to angular distance from the equinoctial. Pèlerin's decision to use *declination* to mean distance from the equinoctial reflects a tendency toward what was becoming standard usage. Cf. *Theoreicae novae planetarum Georgii Puerbachii* (Venice, 1472), "De declinatione et latitudine": "Declinatio stelle est distancia ipsius ab equinoctiali.... Latitudo autem stelle est distancia eius ab ecliptica...."

[96] Cf. M II.35, 36; Chaucer II.30, 35.

[97] *endroit du milieu du zodiaque, c'est endroit de la ecliptique.* The ecliptic line, i.e., the circle traced by the sun's annual course, is a true line (having no breadth), whereas the zodiac is a circular band extending 6° on each side of the ecliptic. Cf.

Chaucer I.21: "... the zodiak in hevene is ymagined to ben a superfice contenyng a latitude of 12 degrees, whereas the remenaunt of cercles in the hevene ben ymagined verrey lynes withoute eny latitude. Amiddes this celestial zodiak is ymagined a lyne which that is clepid the ecliptik lyne, under which lyne is evermo the wey of the sonne." Pèlerin, like Chaucer, is describing the *zodiac in heaven*. On the rete of the astrolabe, by contrast, only the inner or northern part of the zodiacal band is represented, so that the outer edge of the zodiac on the rete is the ecliptic.

[98] *latitude.* Pèlerin here defines latitude in the technical sense of angular "distance from the ecliptic." (See n. 95, above.) But latitude in that sense is irrelevant to the procedure he then describes, the point of which is not distance from the ecliptic but merely, as Chaucer points out, distance from "the wey where as the sonne went thilke day" (II.30, lines 14–15). Messahalla does speak of a planet's having *latitude* ("latitudinem") from the sun's path ("via solis"), and "via solis" does generally mean the ecliptic, as indeed it does elsewhere in Messahalla. See M (Seld) k and cf. Oresme, *Traité de l'espere* (fol. 6v): "la voie du soleil qui est appellee ecliptique" and Chaucer I.21, lines 41–42: "the ecliptik lyne, under which lyne is evermo the wey of the sonne." Here, however, Messahalla refers to distance not from the sun's annual path (the ecliptic) but from its diurnal path (the way of the sun that day).

[99] The procedure is as follows: (1) find the planet's ecliptic longitude, preferably from a table of longitudes; (2) turn the rete so that the degree of the planet's longitude lies on the meridional line and read the altitude of that degree from the almucantars; (3) using the alidade, take the actual altitude of the planet when it is on the meridian; and (4) compare the altitude of the degree, measured in almucantars, with the altitude of the planet, taken with the alidade. Clearly, Pèlerin regards the procedure as yielding latitude from the ecliptic, as *latitude* is defined at the beginning of II.15. It sounds fairly plausible, but it will not work because altitude has reference to the observer's zenith and horizon, and latitude has reference to the zodiacal poles and the ecliptic. See the preceding note.

[100] *4 nuitees ou 6 apres.* M II.36: "post tertiam noctem uel quartam"; Chaucer II.35: "the thridde or fourthe nyght next folewing."

[101] This paragraph, like its counterparts in Messahalla (II.36) and Chaucer (II.35), presents a method that is imperfect because of its failure to take into account the incline of the planets' orbits to the ecliptic. North (*Universe,* 71) remarks that in this matter Chaucer "put more faith in his source than it was worth." So did Pèlerin. The incline, however, was generally held to be negligible (Poulle, *Instruments de la théorie* 1:7).

[102] Cf. M II.37; Chaucer II.36. This chapter is about using an astrolabe to divide the sky into astrological "maisons" in preparation for the casting of a horoscope. Pèlerin's third paragraph translates M II.37; the rest of the chapter is mostly astrological instructions for drawing the horoscope figure. In his astrological treatise, the *Livret de eleccions,* Pèlerin had already translated this same chapter from Messahalla and had added to it even more elaborate astrological explanations. We reproduce the pertinent portions of that treatise below as App. C. Messahalla wrote a second chapter on house-division (M II.38), but Pèlerin did not translate it, although Chaucer did (II.37). The term *maisons* is explained as follows. Astrologically, a planet is said to have certain special powers when it is located in the zodiacal sign designated as its "maison" (domus, house). It is said to have other powers deriving from its location within one or another of the twelve segments into which the sky is divided with respect to the horizon and meridian. These segments are also called "maisons," and it is they that are the subject of the present chapter.

[103] *opposit.* M II.37: "nadayz"; M (Seld) II.37: "nadair"; Chaucer II.36, line 15: "nader"; II.37, line 2: "opposit." See n. 61, above.

¹⁰⁴ M II.38 begins, "Item, habito ascendente et aliis tribus angulis. . . ." Pèlerin in this paragraph (like Chaucer in II.37, lines 1–7) elaborates by explaining how it is that, having found the ascendant according to previous instructions, one has in effect also found the other three "maisons principaulz," otherwise called "angles."

¹⁰⁵ *faire une figure.* I.e., draw a horoscope-chart. See the chart reproduced in App. C, below.

¹⁰⁶ *sur chascun commencement naturel et a chascune heure.* Pèlerin refers here to the principal occasions on which an astrologer would wish to cast a horoscope. Cf. the second paragraph of the excerpt taken from Pèlerin's *Livret* (App. C), where he says that the power of planets and signs is directed by their arrangements "dedens les 12 maisons sur chascun commencement naturel, come a l'eure de revolucions de ans, grans conjunccions, eclipses, nativités, questions, et eleccions. . . ." Cf. also the remarks on horoscopes that Chaucer added to Messahalla: "The ascendent sothly, as well in alle nativities as in questions and eleccions of tymes, is a thing which that these astrologiens gretly observen" (II.4, lines 1–4).

¹⁰⁷ Nothing in Messahalla or Chaucer corresponds exactly to this chapter, but Pèlerin's II.3 (M II.2; Chaucer II.3) describes how one finds the time by taking the altitude of the sun. The present chapter is just a reversal of that procedure: by knowing the time, one finds the sun's altitude.

¹⁰⁸ *en leurs parties du monde.* I.e., to the east or west of the meridian.

¹⁰⁹ Cf. M II.45; Chaucer II.41 (Pèlerin's "Premiere partie"); and M II.46; Chaucer II.42 (Pèlerin's "Seconde partie").

¹¹⁰ *Pour la maniere de ce chapitre,* etc. This introductory paragraph of Pèlerin's is not paralleled in Messahalla or Chaucer.

¹¹¹ *Se nous voulons mesurer aucune chose haute,* etc. The orderly procedure, which Messahalla recommends, is to set the rule at 45° and move from or toward the object until its summit appears through the holes in the vanes of the rule. Then, allowing for the distance from our eyes to the ground, the distance between us and the object is equal to the height of the object. Pèlerin reverses the procedure, saying that we sight the summit by the rule and then, keeping the summit sighted, move backward or forward until the rule falls on 45° ("le dyametre": see the following note). The part of M II.45 to which these lines correspond is omitted by Chaucer.

¹¹² *le dyametre d'un des quarrés.* When the rule is set at 45° in the border, it will pass diagonally through the middle ("dyametre") of one of the two squares formed by the shadow scales. Cf. Raymond, 897: ". . . pone regulam super altitudinem 45 graduum quarte altitudinis scilicet astrolabii per quam solis et stellarum altitudines accipere soles, et hoc facies sic ut regula dividat quadrans per medium."

¹¹³ *adjouster tant comme il a de nostre oeil a terre.* Cf. M II.45: "cum additione stature tue a visu usque ad terram."

¹¹⁴ *avecques nostre hauteur.* Cf. M II.45: "sibi addita stature."

¹¹⁵ *afin que nous puissons dire.* In a sense, we know by looking how high a thing is; we measure and quantify, says Pèlerin, in order that we may say how high it is. The whole paragraph is Pèlerin's addition to his source.

¹¹⁶ *un de costés ou sur l'autre de la quarreure.* The two sides of the square here referred to are the shadow scales, usually distinguished from one another as *umbra extensa* (also called *umbra recta*) and *umbra versa.* Cf. M II.42: "Quadrantis in astrolabio constituti 2 sunt latera. . . . Sed notandum, quod latus inferius uocatur umbra extensa; et aliud latus vmbra uersa. . . ." Cf. also Raymond, 898: "Deinde aspice utrum regula ceciderit super latus umbra recte sive verse." If the object is sighted at an angle greater than 45°, the rule falls on the *umbra extensa* (or *recta*); and if at less than 45°, it falls on *umbra versa.* [Chaucer, I.12, lines 7–9, reverses the terms *umbra recta* and *umbra versa.*]

[117] *parti en 12 parties qui sont appelees pointes.* Cf. M II.42: "in 12· partes equales diuisa, que uocantur puncta umbre"; Chaucer I.12: "the skale ... serveth by his 12 pointes."

[118] *trop pres ... trop loins.* If the rule falls on the *umbra extensa* ("le costé de querreure prouchaine de nous"), then our distance from the object is less than its height ("nous sommes trop pres de tour"). If the rule falls on the *umbra versa*, our distance from the object is greater than its height ("nous sommes trop loins"). Cf. M II.45: "Et si fuerit super umbram extensam, est altitudo maior longitudine; si uero est super uersam, minor longitudine."

[119] *Et pour ce ... et 3, le quart.* The manuscript appears to be garbled here, and we have printed a reconstruction of it. The original, with italics added, reads as follows:

Et pour ce a il encore une autre riulle . car se il nous plaist a mesurer sans ce que *la riulle chiet . et quantes pointes elle trenche . et en quelle porporcion les pointes sont a . 12 . cest a dire la mutel . le tiers* sur le dyametre de la quarreure. Nous devons veoir sur quel costé *la riulle chiet . et quantes pointes elle trenche . et en quelle porporcion les points sont a . 12 . cest a dire la mutel . le tiers* ou le quart . comme . 6 qui sont la muté de . 12 . et . 4 . le tiers et . 3 . le quart.

Our surmise is that the words in italics were mistakenly copied twice. The point to be made in the passage is that either the observer moves and the rule remains set on the diameter, or the observer stays in one place and the rule is moved. If the observer does stay in one place, as Messahalla says (II.45: "ita ut non remoueris a loco vna"), then the rule must be moved. Pèlerin seems to be saying what amounts to the same thing: that if the fixed setting of the rule on the diameter is dispensed with ("sans ce que soit sur le dyametre," i.e., if the rule is moved), then the observer can stay in one place and compute proportions of heights to distances.

[120] *Et tousjours devons adjouster la hauteur de nostre oeil de la terre.* Cf. Raymond, 898: "... semper quantus est ab aspectu tuo in terram superaddi necesse est."

[121] *Et nous devons garder,* etc. These remarks on the quality of instruments are not taken from Messahalla.

[122] *Seconde partie.* This procedure, which follows M II.46, applies when the observer is far enough from his object so that the rule falls on the *umbra versa*. Chaucer II.42, also following M II.46, calls the method a "maner of werkinge, by *umbra versa.*"

[123] *aucune partie proporcionele de 12 poins.* Cf. Chaucer II.41: "even in a point," i.e., exactly on one of the points, rather than somewhere between points. This hint for making calculation easier is not in Messahalla, but it is obvious.

[124] *le nombre combien de fois des pointes.* I.e., the quotient—the number representing how many times the points are contained (M II.46: "continentur") in 12. Cf. Chaucer II.42: "partyes of 12."

[125] *3 ..., le quel nombre est appelé quatre fois.* I.e., three is the number of times that four (the number of points taken at the first station) goes into twelve. Messahalla II.46 uses, as his example, readings of three points at the first station and two points at the second. He calls *four* the "continens tenarij" of the first station, where the number of points at the sighting was *three*. He first writes ".3., que in latere umbre quater continentur" and then, a little below, "continens ternarij, scilicet .4." Note that the numbers in Pèlerin's example are different from Messahalla's.

[126] Cf. M II.47; omitted by Chaucer.

[127] There is no equivalent to this chapter in either Messahalla or Chaucer.

Appendices

Appendix A

In the main, both Pèlerin's *Practique* I and Chaucer's *Astrolabe* I come not from Messahalla's Part I ("De compositione astrolabii") but from a short introduction to his Part II ("De operatione vel utilitate astrolabii"). It is printed in Gunther's edition of Messahalla, 217–18, but for convenience is also reproduced here. Our text is taken from Oxford, Bodleian MS Selden Supra 78, fol. 64r, and where it differs from Gunther's in ways that might conceivably influence a reader or translator, we have noted the differences within brackets. Also, we have expanded the manuscript's abbreviations and have assigned letters to the sub-sections for easier reference.

In the English translation, we adopt the Latin readings of Gunther's edition in items f and n, where his text is clearly superior. The Arabic-derived terms in items a, b, k, and m were never fully assimilated into Western usage. They are followed, in brackets, by the Arabic forms of those terms as given in Morley, *Planispheric Astrolabe*, 9–21.

Incipit practica astrolabii siue rememoracio partium astrolabii.

Nomina instrumentorum sunt hec.

a. Primum est armilla suspensoria ad capiendam altitudinem, et dicitur arabice alhantica [Gunther: alhahucia].

b. Secundum est alhantabor [Gunther: alhabor], id est ansa, que iungitur ei.

c. Postea mater, rotula scilicet, in se continens omnes tabulas cum aranea, cui coniungitur margolabrum scilicet in 360 gradus divisum.

d. Tabule autem ab hac contente signantur [Gunther: figurantur] 3 [Gunther: tribus] circulis quorum minor est circulus cancri, et medius circulus equinoccialis, et maximus circulus capricorni.

e. Postea almucantharath [Gunther: circuli almucantherath] qui sunt circuli in medietate superiori descripti, quorum quidam sunt integri, quidam apparent imperfecti; quibus prior est orizon, dividit enim 2 hemisperia [Gunther: duo emisperia].

f. Centrum autem inferioris [Gunther: interioris] almucantharath cenit capitum nominatur.

g. Deinde sunt [Gunther: est] azimuth, circuli partes almucantharat [Gunther: qui sunt partes circulorum almucantharat] intersecantes.

h. Post quos sunt hore in medietate inferiori descripte.

i. Inter horas uero due sunt crepesculorum linee.

j. Postea linea medii celi, que est descendens ab armilla per centrum in oppositam partem astrolabii. Cuius medietatis a centro in armilla dicitur linea meridiei. Alia dicitur angulus terre et media noctis.

k. Post hec alhancabuz [Gunther: Post hec et segiutur alhantabuth], id est aranea, in qua sunt signa cum zodiaco constituta, stella quoque fixe, in quo via dicitur esse solis; et quiquid fuerit infra motum captis arietis et libre, ex hoc zodiaco dicitur esse septentrionale. Quod autem extra, meridianum dicitur.

l. Postea est almuri, quod ostensor dicitur latine, denticulus scilicet, extra circulum capricorni in alhantabuz [Gunther: alhancabuth] relictus. [Gunther construes "in alhantabuth relictus" with the following clause.]

m. Deinde almehair [Gunther: almenath], id est foramen quod est

Here begin the uses of the astrolabe and a review of its parts.

The names of the instrument are these:

a. First is the suspending ring, used when taking altitudes, and it is called alhantica [*halka*] in Arabic.

b. Second is alhantabor [*habs*], that is the handle to which it is joined.

c. Next is the mother, a sort of wheel which contains within itself the plates and the spider [i.e., rete] and to which is attached a marginal rim divided into 360 degrees.

d. The plates contained in it [i.e., in the mother] are inscribed with three circles, of which the smallest is the circle of Cancer, the middle is the equinoctial, and the largest is the circle of Capricorn.

e. Next are the almucantars, which are circles inscribed in the upper part [of a plate], of which some are complete and others appear incomplete. The first of these divides the two hemispheres.

f. The center of the innermost almucantar is called the zenith over one's head.

g. Next are the azimuths, which are parts of circles intersecting the almucantars.

h. Then there are the hours, inscribed in the lower part [of the plate].

i. Among the hours are the lines of the two crepuscules.

j. Then there is the line of mid-sky, descending from the ring through the center of the astrolabe to its opposite edge. The half [of the line] from the center to the ring is called the meridional line. The other half is called the angle of earth or the line of midnight.

k. After these there is the alhancabuz [*ankabút*], that is the spider, in which are the fixed stars and the signs constituting the zodiac, in which the path of the sun is said to be. And whatever is within the moving of the heads of Aries and Libra is said to be northern from the zodiac. What is outside is called southern.

l. Next is the almuri, which is called *ostensor* in Latin: it is a little tooth remaining outside Capricorn on the alhancabuz.

m. Next is the almehair [*mahan*], the hole that is in the middle of

in medio rethis, in quo est axis retinens tabulas climatum in quam intrat vnus equus [Gunther: intrat alphaeraz, id est equus] restringens araneum cum rotula quasi cuneus.

n. Et in alterea parte matris sunt 2 circuli equacionis solis, quorum vnus continet [Gunther: exterius quorum continet] numerum dierum anni 365 et sub eo scribuntur nomina mensium; et alius circulus graduum signorum [Gunther: alius signorum gradus] infra quem scribuntur nomina signorum.

o. Postea est 4 [Gunther: quarta] capiende altitudinis.

p. Postea quadrans, cuius latera in 12 puncta divisa sunt.

q. Sequitur regula que circumuoluitur in dorso astrolabii, in qua sunt tabule perforate ad capiendum altitudinem solis in die et stellarum in nocte.

the rete. In the hole goes the axis-pin that holds the climate plates, and into the pin goes the horse, as a wedge holding the spider to the wheel [i.e., mother].

n. On the other side of the mother are two circles for locating the sun. The outermost of them contains 365, the number of days in a year, and beneath are written the names of the months. In the other circle are the degrees of the signs, beneath which are the names of the signs.

o. Then there is a quadrant for taking altitudes.

p. Then there is a square, the sides of which are divided into 12 points.

q. Then follows the rule that turns on the back of the astrolabe. On it are perforated plates [i.e., vanes] for taking the altitude of the sun by day and the stars by night.

Appendix B

The manuscript of the *Practique de astralabe,* unlike many manuscript treatises on the astrolabe, provides no figures or diagrams to accompany the text; nor does Pèlerin's text suggest, as Chaucer's does in some copies, that he intended to include visual aids. The omission may be due to his writing for an audience that had astrolabes ready in hand. Charles V is said to have kept, not only at the Hôtel Saint-Pol but at all his residences, at least one astrolabe at his reading table (Fowler, *Plantagenet and Valois,* 189). The library in the Louvre that Charles put at the disposal of his friends was also provided with astrolabes (Labarte, *Inventaire,* item 2427). Figures from various manuscripts of Chaucer's treatise are well reproduced in Fisher, ed., *Complete Poetry and Prose of Geoffrey Chaucer,* and Benson, ed., *Riverside Chaucer.* North, *Chaucer's Universe,* 49, 58–59, reproduces four figures and provides detailed discussions of them. Figures from Messahalla manuscripts are printed in Gunther's *Chaucer and Messahalla,* and pictures of French astrolabes dating from around Pèlerin's time are printed in Gunther's *Astrolabes of the World,* 2:339–343. We here provide figures taken from the copy of Messahalla in Oxford, Bodleian MS Selden Supra 78.

Figure 1, showing a climate plate, is a composite of figures taken from fols. 58r ("almucantharath"), 59r ("azimuth") and 59v ("figura 12 horarum inequalium").

Figure 2, the rete, comes from fol. 57r, where it is labeled with five synonyms: "rethe, volvella, walzagora, alhancabuz, aranea."

Figure 3, the back of the astrolabe ("figura dorsi astrolabii"), comes from fol. 52r.

Figure 4, the alidade, comes from fol. 53r. It is labeled with three synonyms: "allidada, regula, mediclinum." One of the sighting vanes is labeled "pinnula."

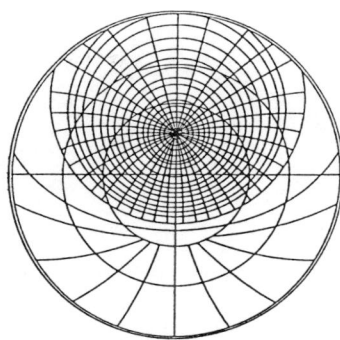

Figure 1. A climate plate ("table"), described in *Practique* I.6–14. Of the circles whose center is the center of the plate, the smallest is the circle of Cancer, the next is the celestial equator ("equinoctial"), and the next is the circle of Capricorn. The outermost circle is the edge of the plate. The vertical line and the horizontal line that cross at the center of the plate form the meridional line, the midnight line, and the east and west lines. The lines radiating from a point (the zenith) in the upper part of the plate are azimuths. The circles surrounding the zenith are almucantars. In the lower part of the plate are lines of unequal hours, radiating from the circle Cancer, and, just above them, the twilight line.

Figure 2. The rete ("reys," "rethe"), described in *Practique* I.15–16. The graduated circular band represents the zodiac. The notched triangular pointer at the top of the zodiac is the almuri. The flame-like pointers ("longuetes") mark the positions of certain fixed stars.

Figure 3. The back ("dos") of the astrolabe. At the very top is a protuberance ("petite hautesse") to which the suspending ring ("armille") would be attached (*Practique* I.2, 3). Just below the center of the back are the graduated shadow scales ("un demy quarré"), numbered by fours up to twelve on each side (*Practique* I.19). Just above the shadow scales are the circles of hours (*Practique* I.19).

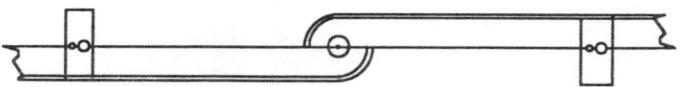

Figure 4. The alidade ("reigle"), described in *Practique* I.20. It pivots from its center on the center of the back. Note the pairs of small peep-holes in the sighting vanes affixed near the ends of the alidade.

On the face of some astrolabes there is another rule, similar to the one on the back but lacking the vanes and used only as a pointer, not for taking altitudes. (See Chaucer, *Astrolabe* II.7.) That rule is not depicted in the Bodleian manuscript, and indeed Morley believes it was a Western invention and was therefore not mentioned in versions of Messahalla closest to the original (*Planispheric Astrolabes*, 15–16). Where Chaucer uses the facial rule as a pointer (II.7), Pèlerin uses the almuri (II.5) and in so doing remains more faithful to Messahalla.

Appendix C

The text below is excerpted from fols. 39r-40v of Pèlerin's *Livret de eleccions,* as preserved in Oxford, St. John's College, MS. 164, immediately preceding his *Practique de astralabe*. The third paragraph of the text and part of the fourth are translated from Messahalla II.37, the source of II.16 of the *Practique*.

From *Livret de eleccions*

Et pour ce[1] convient il savoir la maniere de trouver les commencemens des 12 maisons,[2] affin que a chascune heure quant il est mestier nous puissons trouver leur commencement, et leur commencemens mectre en une figure des 12 parties, affin que nous puissons veoir visiblement et ouvertement leur nombre et quantité et les endrois des planetes dedenz les signes et maisons.

Car toute nature[3] –venant du firmament sur toutes choses subsistans de la mixcion des quatre elemens, de generacion et corrupcion, fortune et infortune, sur chascune creature particuliere–apres dieu vient de la vertu du firmament, des 12 signes et planetes generalment selonc leur mouvement et lieus et regars et figures dedenz les 12 signes et particulierement selonc leur figures et en drescemens et force et foiblesce dedenz les 12 maisons par chascun commencement naturel: come a l'eure de revolucions de ans,[4] grans conjunccions,[5] eclipses, nativités, questions,[6] et eleccions[7] –a chascune racine en parties des jugemens en astrologie. Et nous convient tousjours drescier la figure et enseingnier les commencens des 12 maisons.

Comment nous devons esdrecier a faire [a] chascun temps la figure.

La figure et commencement des 12 maisons nous prenons en 2 manieres.[8] Car aucune fois nous la prenons de la cognoissance de temps passé de une journee non complete,[9] et aucune fois nous prenons le degré ascendant par la hauteur du soleil ou par une estoile fixe,[10] come a nativités & questions. Et aucune fois nous faisons par nostre volenté un degré ascendre selonc la maniere de nostre entencion, et ceste maniere nous usons commencement en la partie des eleccions. Et ces 2 manieres sont applicables en practique par pluseurs practiques. Mais je les mettray et monsterray plus clerement & ouvertement en toutes les manieres que je pourray trouver par le astrelabe.

Comment se treuve le degré ascendant.

Quant aucunes des heures equales sont cogneues par horologe ou par une conjunccion, revolucion, ou autre fait et nous voulons savoir

From *Livret de eleccions*

We must therefore know how to find the beginnings of the 12 houses so that we can find their beginnings whenever we need to do so and put them in a figure of the 12 parts, so that we can clearly and easily see their number, quantity, and the places of the planets in the signs and houses.

For all nature comes from God. It flows from the firmament, over any individual being, over all things consisting of a mixture of the four elements and subject to generation and corruption and good and bad fortune. It comes, after God, from the power of the firmament, the 12 signs, and the planets, generally according to their movement, places, aspects, and figures in the 12 signs and particularly according to their figures and in the direction and force and weakness in the 12 houses at each natural beginning at each root in judicial astrology. The natural beginnings are the times of revolutions of years, great conjunctions, eclipses, nativities, questions, and elections. We must always cast the figure and indicate the beginnings of the 12 houses.

How we must proceed in order to cast a figure at any time.

We take the the figure and beginning of the 12 houses in two ways. In one way, we take it from our knowledge of how much time has passed within an incomplete day, or we take the ascendent degree by the altitude of the sun or a fixed star, as in nativities and questions. In the other way, we arbitrarily choose a degree as ascendent according to our own purposes, and this way we use the beginning when we make elections. And these two ways are applicable in several practices. But I will set them out and exhibit them more clearly and openly in all the ways that I can find relating to the astrolabe.

How the ascendant degree is found.

When the time in equal hours is known by a clock or by a conjunction, revolution, or other event and we want to know the ascendant

le degré ascendant, nous devons mectre le degré du soleil sur la journee sur la ligne de midi en astrelabe et considerer le almuri, en quel endroit il touche le limbe. Et devons tourner[a] le reche pour chascune heure 15 degrés a tant de fois come nous avons des heures. Et quel degré du zodiaque touche le premier almucanterath devers orient, il sera ascendant au bout des dites heures. Mais se nous prenons la hauteur du soleil et voulons savoir le degré ascendant, nous considerons la hauteur du soleil ou la hauteur d'un autre estoile par la volvelle de astrelabe[11] et considerons se le soleil ou l'estoile a passé midi ou se elle est d'entre midi & orient. Et a celle partie en la quelle le soleil est ou l'estoile est nous compterons tant de degréz des almucantarath de la partie de orient ou de occident et mectons droitement le lieu du soleil ou de l'estoile sur la hauteur du almucantarath. Et quel des signes et son degré touche le premier almucantarath devers orient et est ascendant en l'eure de nostre consideracion.

Et se nous faisons de nostre volenté un ascendant, nous devons tourner le reche de astrelabe jusques a tant que le degré que nous voulons estre ascendant soit sur le almucantarath premier orientel.

Comment nous devons trouver le commencement des 12 maisons.

Quant le degré ascendant est trouvé par aucune de ces 2 manieres et nous voulons trouver les commencemens des autres maisons, donques nous devons situer le dit degré ascendant en astrelabe sur le orison de la table qui desert a nostre region. Et donques le signe et degré qui touche la ligne de midy commence la 10e maison, et qui touche le orison devers occident la 7e. Et quant el touche la ligne de minuit, commence la 4e maison.[12]

En exemple, se le 10e degré de Aries est ascendant et est mis sur le orient nous trouverons en midi le quart degré de Capricorne, et en occident le 10e de la Libre, et a minuit le quart de Cancre, en la table de region de Paris en France.

Et donques nous dresserons une figure la quelle nous partirons en 12 parties. Et escrivons le degré ascendant en milieu de la dite figure devers la main senestre et le 10e du Libre en son opposit et le quart de Capricorne a midi, comme monstre ceste figure.

degree, we must put the degree of the sun that day on the meridional line in the astrolabe and determine in which place the almuri touches the limb. Then we have to turn the rete 15 degrees for each hour as many times as we have hours. The degree of the zodiac which touches the first almucantar toward the east side will be ascendant at the end of these hours. But if we take the altitude of the sun and wish to known the ascendant degree, we consider the altitude of the sun or the altitude of another star by the volvelle of the astrolabe and consider also whether the sun or star has passed the meridian or is between the meridian and the east. Toward the side where the sun or star is, we count the appropriate number of degrees in almucantars, on the east side or the west side, and place the longitude of the sun or the star directly on the almucantar of its altitude. The sign and degree which touches the first almucantar toward the east is ascendent at the time under consideration.

If we choose an ascendant arbitrarily, we must turn the rete of the astrolabe just far enough so that the degree that we want to be ascendant is on the first almucantar on the east side.

How we are to find the beginning of the 12 houses.

When the ascendant degree is found by one of the two methods and we wish to find the beginnings of the other houses, we have to situate the said ascendant degree in the astrolabe on the horizon of the plate which represents our region. The sign and degree which touches the line of midday begins the 10th house and that which touches the horizon toward the west, the 7th. When it touches the line of midnight, it begins the 4th house.

For example, if the 10th degree of Aries is ascendant and is put on the east, we will find the 4th degree of Capricorn on the meridian, the 10th of Libra in the west, and the 4th of Cancer on the line of midnight, on the plate for the region of Paris in France.

We draw a figure which we divide into 12 parts. We write the ascendant degree in the middle of this figure toward the left hand, the 10th of Libra in its opposite, and the fourth of Capricorn at the meridian, as this figure shows.

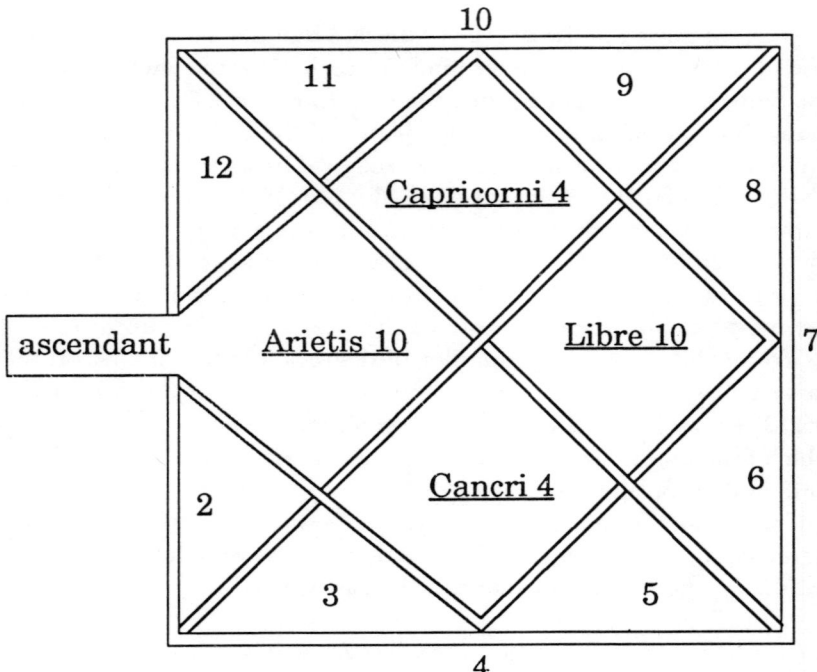

Depuis que lé 4 souverains maisons[b] sont trouvés,[13] il faut apres trouver les autres 8. Nous devons mectre le degré ascendant sur le bout de la 8[e] heure de l'astrelabe. Et donques le signe et son degré qui cherra sur la ligne de minuit commencera la seconde maison, le quel signe et son degré nous escrivons en la figure au lieu de la 2e maison et son opposit signe et degréz dedenz la 8e maison. Apres, nous mectons le degré ascendant sus le bout de la 10e heure, et le signe et son degré qui cherra sur la ligne de minuit sera commençant et escript dedenz la tierce maison. Encore faut il trouver le commencement seulement de 4 maisons. Pour quoy nous mectrons le opposit signe et le degré de ascendant sur le bout de la seconde heure, et le signe et son degré cheant sur la ligne de minuit commencera la 5e maison, et son opposit la 11e. Et encore nous mectrons le opposit de ascendant sur le bout de la quarte heure, et le signe et son degré touchant la ligne de minuit commence la 6e maison, et son opposit la 12e.

Et en telle maniere nous dressons par le astrelabe a chascun temps nostre figure. Et ceste operacion je ai mis en exemple et dressee la figure sur le 10e degré de Aries en ascendant afin que ceste oevre se puisse ouvertement entendre.

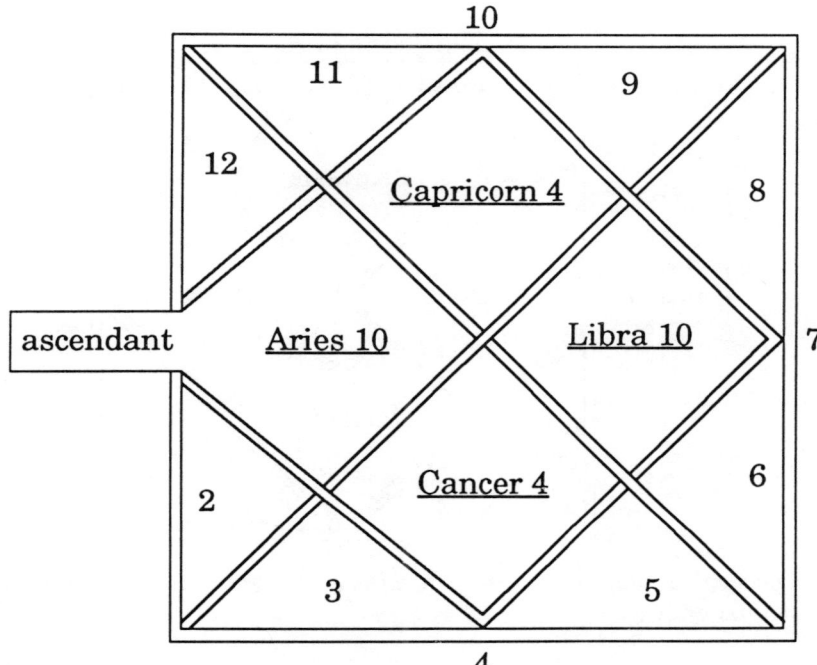

After the four principal houses are found, we have to find the other eight. We have to put the ascendant degree on the end of the 8th hour line of the astrolabe. The sign and degree which falls on the line of midnight begins the second house, which we write in the place of the 2nd house in the figure and its opposite sign and degree in the 8th house. Then we put the ascendant degree on the end of the 10th hour line, and the sign and degree which falls on the line of midnight is the beginning and is written in the third house. Then we have to find the beginning of only four houses. It is for this reason we put the opposite sign and degree of the ascendant on the end of the second hour line, and the sign and degree falling on the line of midnight begins the 5th house, and its opposite the 11th. Then we put the opposite of the ascendant on the end of the fourth hour line, and the sign and degree touching the line of midnight begins the 6th house and its opposite the 12th.

In this way we cast our figure by the astrolabe for any time. To show clearly how this work should be understood, I have shown this operation as an example and based the figure on the 10th degree of Aries ascending.

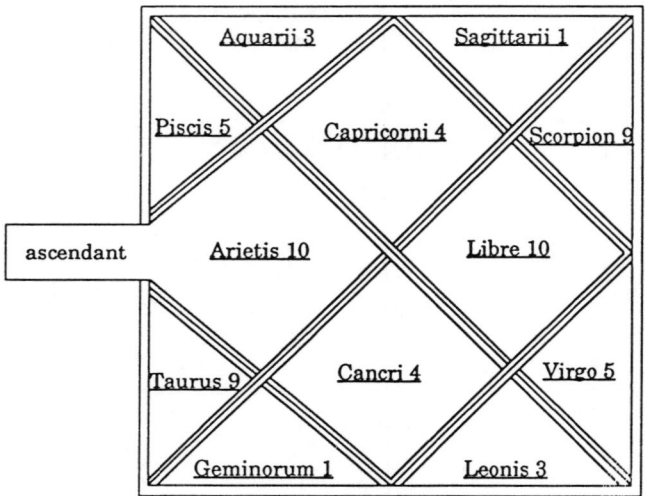

Comment les planetes dovient estre mises en la figure, la quelle figure est ci-apres dressee.

Quant la figure a chascune racine des jugemens est dressee et les commencemens des 12 maisons assignees, donques convient mectre les planetes 7 et le chief de dragon et la keue[14] dedenz la figure, chascune en telle maniere comme sera le signe qui tient le planete dedenz le almenach sur la journee sur quoy nostre racine et nostre figure est dressee. Vez ci l'exemple et patron de la figure.

Et pour ceste raison, miex entendre et sans deffaute mectre en oevre, nous devons regarder de chascune maison de quel signe et degré elle commence, adjungant 5 degréz[15] au commencement. Et apres devons regarder la maison qui vient apres, du quel signe et degré elle commence en laissant 5 degrés. Et se aucune planete est trouvee par le almanach entre les 2 termes, elle doit estre escript en la dite maison.

En exemple, la premiere maison a ceste figure commence de 10e degré de Aries, au quel je adjunge 5 degréz, si que elle sera commençant en sa vertu de 5 degrés de Aries. Et la seconde maison commence de 9 degrés de Taurus, au quel je adjunge 5 degrés, et elle sera commençant de 4 de Taurus.

Et pour ce se je trouvoie aucune des planetes d'entre les 5 degréz de Aries jusques a 4 degréz de Taurus, le dit planete doit estre escript dedenz la premiere maison. Et en tele maniere de toutes les autres maisons devons nous faire.

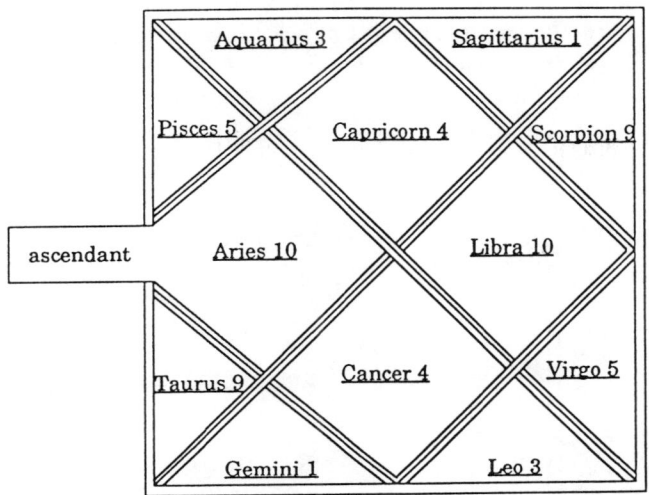

How the planets must be put in the figure, as drawn below.

When the figure for each root of any judgment is cast and the beginnings of the 12 houses assigned, we have to put the seven planets and the head and tail of the dragon in the figure, each in the sign in which the almanach locates it on the day on which our root and our figure is cast. See the example and model figure below.

To better understand and to accurately do the work, we have to note at which sign and degree each house begins, adding 5 degrees to the beginning. Then we have to look at the house which comes afterward, adding 5 degrees to the sign and degree at which it begins. If a planet is found by the almanach between the two terms, it should be written in the said house.

For example, the first house in this figure begins at the 10th degree of Aries, to which I add 5 degrees so that its power will begin at 5 degrees Aries. The second house begins at 9 degrees Taurus, to which I add 5 degrees, and it will be begin at 4 degrees Taurus.

Thus, if I find one of the planets between 5 degrees of Aries and 4 degrees of Taurus, the said planet should be written in the first house. We should do all the other houses in this way.

Et pour general exemple[16] je ay mise chascune planete a la figure yci faite en quelle maison elle estoit selonc la dite figure l'an de grace 1360, en la derreniere journee de octobre. C'est ci la figure toute faite eu commencement des maisons et lieus des planete[s].

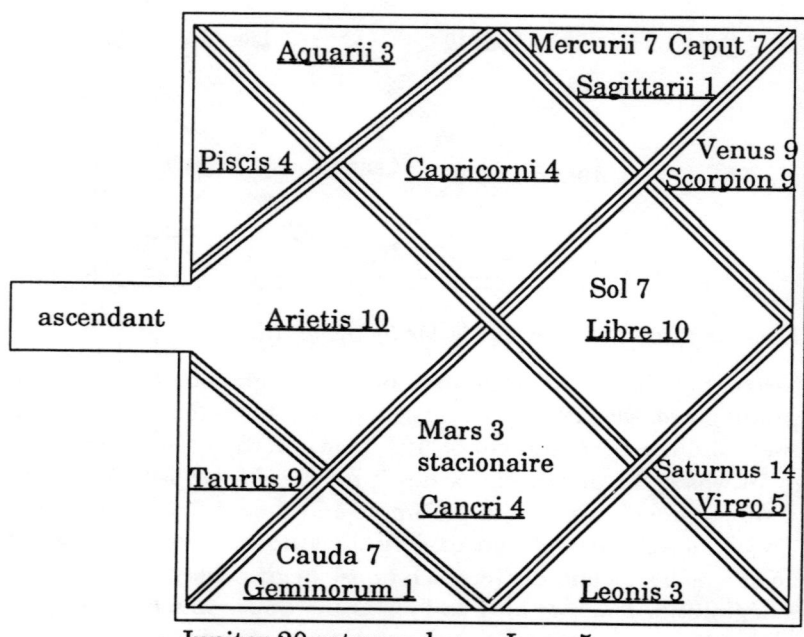

As a general example, I have put each planet in the figure below in the house in which it belongs according to the figure for the last day of October, the year of our Lord 1360. Here is the completed figure of the beginning of the houses and the places of the planets.

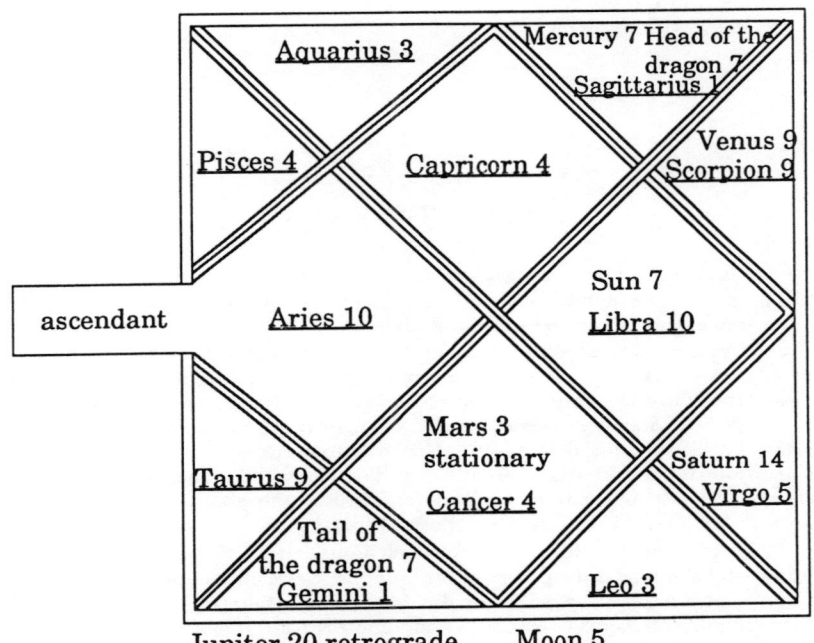

Textual Notes

^atourner. MS: trouver.
^bmaisons. MS: planetes.

Explanatory Notes

[1] *Et pour ce* etc. It is necessary to know how to find the beginnings of houses at each moment because the houses change from moment to moment as each degree of the zodiac rises above the eastern horizon.

[2] *12 maisons*. The kind of astrological *maisons* or *houses* Pèlerin here speaks of are segments of sky supposed to have certain significant properties. The astrologer divided the entire sky, both above and beneath the earth, in something like the way an orange is divided into segments. He begins the first segment (*maison*) at his eastern horizon and numbers counter-clockwise to twelve so that the seventh begins at his western horizon, the tenth at the mid-sky above him, and the fourth at the mid-sky beneath. On various systems for dividing the sky, see J. C. Eade, *The Forgotten Sky* (Oxford: Oxford Univ. Press, 1984), 41–47.

[3] *Car toute nature* etc. God is the First Cause of all things, but in terms of natural philosophy it is the power of the immutable sky and of unchanging celestial motion that produces change in earthly things, which, because they are composed of the mutable elements (earth, water, air, and fire), are subject to generation and decay, fortune and misfortune. Pèlerin's words reflect a standard synthesis of astrology with Aristotelian cosmology and element-theory, together with an allusion to the text on which the synthesis is based, Aristotle's *De generatione et corruptione*. Cf. Averroes: "Since ... the eternal circular motion is the motion of the sky, it is manifest that the motion of things subject to generation and corruption comes from that motion and is caused by it" (*Commentarium medivm*, ed. F. H. Fobes [Cambridge, Mass.: Mediaeval Academy of America, 1956], 160, our translation); also, the anonymous thirteenth-century *Auctoritates Aristotelis*, citing *De gen. et corrup.* : "The movement of the sun and the other planets in the oblique circle is the cause of generation and corruption of lower things" (quoted in Tester, *History*, 160). Also, perhaps the most famous citation, Sacrobosco, *De spera*: "By Aristotle in *On Generation and Corruption* [the ecliptic line] is called the 'oblique circle,' where he says that, according to the access and recess of the sun in the oblique circle, are produced generations and corruptions in things below" (Thorndike, *The Sphere of Sacrobosco*, 124). J. D. North remarks that "most of those who accepted an Aristotelian cosmology found themselves obliged to accept astrology as a concomitant" (*Richard of Wallingford* [Oxford: Clarendon Press, 1976], 2:84). The whole synthesis, as decisively achieved by Albumasar, is described by Lemay, *Abu Ma'Shar*, 55–85.

[4] *revolucions de ans*. Entries of the sun into Aries (at each vernal equinox), at

which times horoscopes were cast to make predictions for the coming year.

[5] *grans conjunccions.* Conjunctions of the "superior" planets Saturn and Jupiter, supposed to have long-term effects.

[6] *questions.* Also called *interrogations,* and based on the notion that the horoscope, at the moment an astrologer is asked a question, determines the correct answer.

[7] *eleccions.* Choosing, according to horoscopes, the best times for initiating actions. This is the main subject of the *Livret,* and in the copy of it in the St. John's manuscript, attention is drawn to the following elections by the word "nota" written in the margin: conceiving a child (fol. 64v), lying with a woman for pleasure (68v), gaming for profit (89v), constructing engines for destroying fortresses (92v), and going on a long journey (95v).

[8] *2 manieres.* Cf. *Practique* II.3. One may set the rete of an astrolabe by either of two procedures for different purposes. One is to determine the present state of the heavens and set the rete to represent the sky at that time. The other is to set the rete arbitrarily and then determine the time when the sky will match that arbitrary setting. The first procedure is suited to astrological nativities, since the time of a birth is pretty much beyond one's control and the astrologer wishes merely to know the state of the heavens at that time. The second procedure is suited to elections, since the astrologer knows what conditions of the heavens will be propitious for an action one wishes to take and needs to know when those conditions will be met. These purely astrological considerations help explain, perhaps, why it is that whereas Messahalla in II.1 speaks only of calculations for the present day ("diem presentis"), Pèlerin, *Practique* II.1, speaks of the "journee presente ou autre."

[9] *Car aucune fois ... journee non complete.* Cf. *Practique* II.6.

[10] *et aucune fois ... estoile fixe.* Cf. *Practique* II.2, 3.

[11] *volvelle de astralabe.* I.e., the rete, called "volvella" in the drawing of a rete in Oxford, Bodleian MS Selden Supra 78, fol. 57r (see App. B, above). The place of the sun or a star as represented on the rete is to be set with reference to the almucantars and meridian on the climate plate.

[12] *Quant le degré ascendant est trouvé* etc. With this paragraph, cf. the first paragraph of *Practique* II.16.

[13] *Depuis que lé 4 souverains maisons sont trouvés* etc. Cf. Messahalla II.37. These *"maisons"* are usually called "angles," which is what Pèlerin calls them in the Prologue to the *Practique,* as does Chaucer in the *Astrolabe* II.4, line 46.

[14] *le chief de dragon et la keue.* These are the lunar nodes, the points where the moon crosses the ecliptic moving north (*chief*) and south (*keue*).

[15] *adjungant 5 degrez,* etc. Pèlerin here gives what appears to be a fuller version of a doctrine to which Chaucer refers in *Astrolabe* II.4: "For, after the statutes of astrologiens, what celestial body that is 5 degrees above thilke degre that ascendith, or withinne that nombre, that is to seyn neer the degre that ascendith, yit rekne they thilke planete in the ascendent. And what planete that is under thilke degre that ascendith the space of 25 degres, yit seyn thei that thilke planete is 'like to him that is the hous of the ascendent.' " There is a serious disagreement over what Chaucer's brief reference means (See Chauncey Wood, *Chaucer and the Country of the Stars* [Princeton: Princeton Univ. Press, 1970], 16–17; Eade, *Forgotten Sky,* 46–47; and North, *Chaucer's Universe,* 66–67). Pèlerin's version, however, is clear enough.

[16] *Et pour general example* etc. As Pèlerin says in the *Practique* I.1, planetary positions are most surely learned by consulting an "almanach," that is, a planetary table. By Pèlerin's time the Alphonsine Tables had made all others virtually obsolete (E. Poulle, *Les Tables Alphonsines* [Paris: Centre National de la Recherche Scientifique, 1984], 1). Using the Alphonsine parameters as a base for a microcomputer program,

our colleague Donald W. Olson has determined that Pèlerin's planetary locations for October 31, 1360, are accurate in all cases but one, the location of the sun, which should be 17° Scorpio instead of 7° Libra. It is possible that, in a draft or rough sketch of the horoscope figure, "Sol 17" was written so that it ran into the line dividing Libra from Scorpio. Thus "17" was obscured to look like "7," and the entry was misinterpreted as belonging in Libra. In any case the error is more likely to be scribal than computational, for the positions given in the manuscript place Mercury and the sun 60° apart, which, as Pèlerin would have known, is impossible.

Sigmund Eisner has pointed out to us that the *Kalendarium* of Nicholas of Lynn (cited by Chaucer in the Prologue to his *Astrolabe*) yields information on the equation of houses very similar to that found in Pèlerin. The differences are as follows:

House	Pèlerin	Nicholas
I	Aries 10	Aries 10
II	Tau. 9	Tau. 10
III	Gem. 1	Gem. 8
IV	Can. 4	Can. 4
V	Leo 3	Leo 4
VI	Vir. 5	Vir. 6
VII	Lib. 10	Lib. 10
VIII	Sco. 9	Sco. 10
IX	Sag. 1	Sag. 8
X	Cap. 4	Cap. 4
XI	Aqu. 3	Aqu. 4
XII	Pis. 5	Pis. 6

The data from the *Kalendarium* are found on 164–75 of Eisner's edition.

Bibliography

Bibliography

Abraham ibn Ezra. *The Beginning of Wisdom*. Edited by Raphael Levy and Francisco Cantera. The Johns Hopkins Studies in Romance Literatures and Languages, Extra vol. 14. Baltimore, 1939.

Alchabitius (Al-Qabisi, al-Kabisi). *Introductorius ad iudicia astrologiae*. Oxford, Bodleian MS Selden Supra 78, fols. 25-50; Oxford, St. John's College MS. 164, fols. 119-160 (French); Cambridge, Trinity College MS. 0.5.26, fols. 1-27 (English). See also John of Saxony.

Allen, R. H. *Star-Names and Their Meanings*. New York: Stechart, 1899.

Averrois Cordvbensis. *Commentarivm medivm in Aristotelis De generatione et corrvptione libros*. Edited by F. H. Fobes. Cambridge, Mass.: Mediaeval Academy of America, 1956.

Bate, Henry. *Magistralis compositio astrolabii*. Printed in Gunther, 1932.

Benjamin, Francis S., and G. J. Toomer. *Companus of Novara and Medieval Planetary Theory: Theorica planetarum Edited with an Introduction, English Translation, and Commentary*. Madison: Univ. of Wisconsin Press, 1971.

Benoît, Paul. "Recherches sur le vocabulaire des opérations élémentaires dans les arithmétiques en langue française de la fin du moyen âge." *Documents pour l'histoire du vocabulaire scientifique* 7 (1985): 77-95.

Benson, Larry D., ed. *The Riverside Chaucer*. 3rd ed. Boston: Houghton Mifflin Co., 1987.

Borst, Arno. *Astrolab und Klosterreform*. Sitzungsberichte der Heidelberger Akademie der Wissenschaften, Philosophische-historische Klasse, Jg. 1989, Bericht 1. Heidelberg: Carl Winter Universitätsverlag, 1989.

Brunot, Ferdinand. *Histoire de la langue française*. Tome I, *De L'Epoque latine à la renaissance*. Paris: Librairie Armand Colin, 1966.

Campanus of Novara. See Benjamin and Toomer.

Carmody, Francis J., ed. *Leopold of Austria: "Li Compilacions de le science des estoilles."* Univ. of California Publications in Modern Philology 33. Berkeley, 1947.

———, ed. *Theorica planetarum Gerardi*. Berkeley, 1947.

———. *Arabic Astronomical and Astrological Sciences in Latin Translation: A Critical Bibliography*. Berkeley: Univ. of California Press, 1956.

———, ed. *Astronomical Works of Thabit b. Qurra*. Berkeley and Los Angeles: Univ. of California Press, 1960.

Catalogue des manuscrits de la Bibliothèque de l'Arsenal. 9 t. Paris: E. Plon et cie., 1885-1899.

Christine de Pisan. *Fais et bonnes meurs du sage roy Charles V*. Edited by S. Solente. 2 vols. Paris: H. Champion, 1936-1940.

———. *Le Livre de paix*. Edited by Charity Cannon Willard. The Hague: Mouton and Co., 1958.

Coopland, G. W. *Nicole Oresme and the Astrologers: A Study of His Livre de Divinacions*. Cambridge: Harvard Univ. Press, 1952.

Coxe, Henry O. *Catalogus codicum MSS. qui in collegiis aulisque Oxoniensibus hodie adservatur*. Vol. 2, pt. 6, *Catalogus codicum MSS. Collegii S. Johannis Baptistae*. Oxonii, 1852.

Dedeck-Héry, V. L., ed. "Boethius' De Consolatione by Jean de Meun." *Mediaeval Studies* 14 (1952): 165-275.

Delachenal, Robert. "La Date de la naissance de Charles V." *Bibliothèque de l'Ecole des Chartes* 64 (1903): 94-98.

———. *Histoire de Charles V*. 5 vols. Paris: Librairie Alphonse Picard et Fils, 1909-1931.

———. "Note sur un manuscrit de la bibliothèque de Charles V." *Bibliothèque de l'Ecole des Chartes* 71 (1910): 33-38.

Delisle, Léopold. *Recherches sur la librairie de Charles V*. 2 vols. Paris, 1907.

de Phares. See Simon de Phares.

Dictionary of the Middle Ages. Edited by Joseph L. Strayer. 13 vols. New York: Charles Scribner's Sons, 1982-1989.

Dictionary of Scientific Biography. Edited by C. C. Gillispie. New York: Charles Scribner's Sons, 1970-1980.

Eade, J. C., *The Forgotten Sky: A Guide to Astrology in English Literature*. Oxford: Oxford Univ. Press, 1984.

Elliot, Ralph W. V. *Chaucer's English*. London: André Deutsch, 1974.

Equatorie of the Planetis. See Price.

Les Fastes du gothique, le siècle de Charles V. Paris: Editions de la Réunion des musées nationaux, 1981.

Fine, Oronce. *Theorique des cielz*. Paris, 1528.

Fisher, John H., ed. *The Complete Poetry and Prose of Geoffrey Chaucer*. 2d ed. New York: Holt, Rinehart and Winston, Inc., 1989.

Goldschmidt, E. Ph. *Medieval Texts and Their First Appearance in Print*. New York: Biblo and Tannen, 1969.

Grant, Edward, ed. and trans. *Nicole Oresme:* De proportionibus proportionum *and* Ad pauca respicientes. Madison: Univ. of Wisconsin Press, 1966.

Green, Richard Firth. *Poets and Princepleasers.* Toronto: Univ. of Toronto Press, 1980.

Grosseteste, Robert. *De sphaera.* In *Die philosophischen Werke des Robert Grosseteste, Bischofs von Lincoln,* 10–32. Edited by Ludwig Baur. Münster: Aschendorff, 1912.

Gunther, R. T. *Chaucer and Messahalla on the Astrolabe.* Early Science in Oxford, vol. 5. Oxford: Oxford Univ. Press, 1929.

———. *The Astrolabes of the World.* 2 vols. Oxford: Oxford Univ. Press, 1932.

Hartner, Willy. "The Principle and Use of the Astrolabe." *Oriens-Occidens,* 287–311. Hildesheim: Georg Olms Verlagsbuchhandlung, 1968. Reprinted from *A Survey of Persian Art,* ed. Arthur Upham Pope, vol. 3. Oxford: Oxford Univ. Press, 1939.

Hermannus Contractus. *De mensura et utilitatibus astrolabii.* Printed in Gunther, 1932.

Inventaire du mobilier de Charles V, roi de France. Edited by Jules Labarte. Collection de documents inédits sur l'histoire de France, troisième série, archéologie. Paris, 1879.

John of Sacrobosco. See Thorndike.

John of Saxony. Commentary on Alchabitius, printed with Alchabitius, *Libellus Ysagogicus Abdilazi.* . . . Venice: J. and G. de Gregorius, 1491.

Jordanus de Nemore. See Thomson.

Jourdain, Charles. "Nicholas Oresme et les astrologues de la cour de Charles V." *Revue des questions historiques* 18 (1875): 136–59.

Kren, Claudia. "Homocentric Astronomy in the Latin West: The *De reprobatione ecentricorum et epiciclorum* of Henry of Hesse," *Isis* 59 (1985): 269–81.

Kunitzsch, Paul. "On the Authenticity of the Treatise on the Composition and Use of the Astrolabe Ascribed to Messehalla." *Archives internationales d'histoire des sciences* 31 (1981): 42–62.

———. "Glossar der arabischen Fachausdrücke in der mittelalterlichen europäischen Astrolabliteratur." *Nachrichten der Akademie der Wissenschaften in Göttingen,* Philologische-historische Klasse, 11 (1983): 455–571.

Labarte, Jules. See *Inventaire du mobilier.*

Laird, Edgar. "Astrology in the Court of Charles V of France, as Reflected in Oxford, St. John's College, MS 164." *Manuscripta* 34 (1990): 167–76.

———. "Robert Grosseteste, Albumasar, and Medieval Tidal Theory." *Isis* 81 (1990): 684–94.

Lattin, Harriet Pratt. "Lupitus Barchionensis." *Speculum* 7 (1932): 58–64.

Lemay, Richard. *Abu Ma'Shar and Latin Aristotelianism in the Twelfth Century.* American Univ. of Beirut Publications of the Faculty of Arts and Sciences. Oriental Ser. 88. Beirut, 1962.

———. "The Teaching of Astronomy in Medieval Universities, Principally at Paris in the Fourteenth Century." *Manuscripta* 20 (1976): 197–217.

Leopold of Austria. See Carmody, 1947.

La Librairie de Charles V. Paris: Bibliothèque Nationale, 1968.

Lindberg, David D. *The Beginnings of Western Science.* Chicago: Univ. of Chicago Press, 1992.

Llobet of Barcelona, et al. Treatises printed in Millás Vallicrosa.

Lorch, Richard. "The *sphaera solida* and Related Instruments." *Centaurus* 24 (1980): 153–61.

Lutz, Cora E. *Essays on Manuscripts and Rare Books.* New York: Archon Books, 1975.

Martinez-Duenas-Espejo, Jose-Luis. "La prosa cientifica de Geoffrey Chaucer: Estudio textual y gramatical de *A Treatise on the Astrolabe.*" In *Estudios literarios ingleses: Edad Media,* edited by J. -F. Galvan-Reula, 121–37. Madrid: Catedra, 1985.

Masi, Michael. "Chaucer, Messahalla and Bodleian Selden Supra 78." *Manuscripta* 19 (1975): 37–47.

McAlindon, T. "Cosmology, Contrariety, and the Knight's Tale." *Medium Aevum* 55 (1986): 47–48.

Menut, Albert D. "A Provisional Bibliography of Oresme's Writings." *Mediaeval Studies* 28 (1966): 279–99.

Menut, Albert D., and Alexander J. Denomy, eds. and trans. *Nicole Oresme: Le Livre du ciel et du monde.* Madison: Univ. of Wisconsin Press, 1968.

Millás Vallicrosa, J. M. *Assaig d'Historia de les Idees Fisiques i Matemàtiques a la Catalunya Medieval.* Barcelona: Institució Patxot, 1931. "Apèndix," 272–335: edition of Llobet treatises.

Molenaer, Samuel Paul, ed. *Li Livres du gouvernement des rois: A XIIIth Century French Version of Egidio Colonna's De Regimine principum.* New York: Columbia Univ. Press, 1899. Reprint. New York: AMS Press, 1966.

Morley, William H. *Description of a Planispheric Astrolabe Constructed for Sháh Sultán Husain Safawí, King of Persia, and Now Preserved in the British Museum.* London: Williams and Norgate, 1856. Reprinted in Gunther, 1932.

Neugebauer, O. "The Early History of the Astrolabe." *Isis* 40 (1949): 240–56.

Newe Theorike of Planetis. Cambridge, Trinity College MS. O.5.26, fols. 125–181.

Nicholas of Lynn. *The Kalendarium of Nicholas of Lynn.* Edited by Sigmund Eisner. Athens, Georgia: Univ. of Georgia Press, 1980.

North, J. D. "The Astrolabe." *Scientific American* 230 (Jan. 1974): 96–106.

———, ed. and trans. *Richard of Wallingford. An Edition of His Writings, with Translation and Commentary.* 3 vols. Oxford: Clarendon Press, 1976.

———. *Horoscopes and History.* London: The Warburg Institute, 1986.

———. *Chaucer's Universe.* Oxford: Oxford Univ. Press, 1988.

Oresme, Nicole. *Le Livre du ciel et du monde.* See Menut and Denomy.

———. *Le Livre de divinacions.* See Coopland.

———. *Traité de l'espere.* Oxford, St. John's College, MS. 164, fols. 1–32.

———. *Ad pauca respicientes.* See Grant.

———. *De proportionibus proportionum.* See Grant.

Osborne, Marijane, "The Squire's 'Steed of Brass' as *Astrolabe:* Some Implications of the *Canterbury Tales.*" In *Hermeneutics and Medieval Culture,* edited by Patrick J. Gallacher, 121–31. Albany: SUNY Press, 1989.

Ovitt, George, Jr. "History, Technical Style and Chaucer's *Treatise on the Astrolabe.*" In *Creativity and the Imagination: Case Studies from the Classical Age to the Twentieth Century,* edited by Mark Amsler, 24–58. Newark: Univ. of Delaware Press, 1987.

Patterson, Lee. *Chaucer and the Subject of History.* Madison: Univ. of Wisconsin Press, 1991.

Peck, Russell A. *Chaucer's* Romaunt of the Rose *and* Boece, Treatise on the Astrolabe, Equatorie of the Planetis, Lost Works and Chaucerian Apocrypha: An Annotated Bibliography, 1900 to 1985. Toronto: Univ. of Toronto Press, 1988.

Pèlerin de Prusse. *Livret de eleccions.* Oxford, St. John's College, MS. 164, fols. 33–111.

The Planispheric Astrolabe. Greenwich: National Maritime Museum, 1979. Reprint 1989.

Poulle, Emmanuel. "Le Quadrant nouveau médiéval." *Journal des savants* avril-juin (1964): 148–67, 182–214.

———, ed. "Le Traité d'astrolabe de Raymond de Marseille." *Studi medievali,* 3rd ser., 5 (1964): 866–909.

———. "Horoscopes princiers des XIVe et XVe siècles." *Bulletin de la Société Nationale des Antiquaires de France* (Feb., 1969): 63–69.

———. *Les Instruments de la théorie des planètes selon Ptolémée: equatoires et*

horologerie planétaire du XIIIe au XVIe siècle. 2 t. Geneva: Librairie Droz, 1980.

———. *Les Instruments astronomiques du moyen âge.* Paris: A. Brieux–E. Poulle, 1983.

———, ed. and trans. *Les Tables Alphonsines avec les canons de Jean de Saxe.* Paris: Centre National de la Recherche Scientifique, 1984.

Price, D. J., ed. *The Equatorie of the Planetis.* Cambridge: Cambridge Univ. Press, 1955.

Puerbach, George. *Theorica novae planetarum Georgii Puerbachii astronomici celebratissimi de sale.* Venice, c. 1472. Repr. in *Regiomontari opera collectanea.* Edited by Felix Schmeidler. Onasbrück: Otto Zeller, 1972.

Raymond of Marseilles. See Poulle, 1964.

De Regimine principum. See Molenaer.

Richard of Wallingford. See North, 1976.

Robinson, Pamela. "Geoffrey Chaucer and the *Equatorie of the Planetis:* the State of the Problem." *Chaucer Review* 26 (1991): 17–30.

Sarton, George. *Introduction to the History of Science.* 3 vols. Baltimore: Carnegie Institute, 1927–1948.

Saunders, Harold N. *All the Astrolabes.* Oxford: Senecio Publishing, 1984.

Sherman, Claire Richter. "Representations of Charles V of France (1338–1380) as a Wise Ruler." *Medievalia et Humanistica,* n.s. 2 (1971): 83–96.

Simon de Phares. *Recueil des plus célèbres astrologues et quelques hommes doctes.* Edited by Ernest Wickersheimer. Paris: H. Champion, 1929.

Skeat, W. W., ed. *A Treatise on the Astrolabe; Addressed to His Son Lowys by Geoffrey Chaucer, AD 1391.* Oxford: Oxford Univ. Press, 1872.

———, ed. *The Complete Works of Geoffrey Chaucer.* 7 vols. Oxford: Clarendon Press, 1894–1897.

Southern, R. W. *Robert Grosseteste: The Growth of an English Mind in Medieval Europe.* Oxford: Clarendon Press, 1986.

Steinschneider, Moritz. *Die europäischen Übersetzungen aus dem Arabischen bis Mitte des 17 Jahrhunderts.* 2 vols. in 1. Wein: C. Gerolds Sohn, 1904–1905. Reprint. Graz: Akademische Druck–u. Verlagsanstalt, 1956.

———. *Die hebräischen Übersetzungen des Mittelalters und die Juden als Dolmetscher.* 2 vols. Berlin: Kommissionsverlag des Bibliographischen Bureaus, 1893. Reprint. Graz: Akademische Druck–u. Verlagsanstalt, 1956.

Tester, Jim. *A History of Western Astrology.* New York: Ballantine Books, 1987.

Thabit ibn Qurra. See Carmody.
Theorica planetarum Gerardi. See Carmody.
Thomson, Ron B., ed. and trans. *Jordanus de Nemore and the Mathematics of Astrolabes: De plana spera.* Toronto: Pontifical Institute of Mediaeval Studies, 1978.
Thorndike, Lynn. *A History of Magic and Experimental Science.* 8 vols. New York: Columbia Univ. Press, 1923–1958.
———. *University Records and Life in the Middle Ages.* New York: Columbia Univ. Press, 1944. Reprint. 1971.
———, ed. and trans. *The "Sphere" of Sacrobosco and Its Commentators.* Chicago: Univ. of Chicago Press, 1949.
———. "The True Place of Astrology in the History of Science." *Isis* 46 (1955): 273–78.
———. "The Study of Mathematics and Astronomy in the Thirteenth and Fourteenth Centuries as Illustrated by Three Manuscripts." *Scripta Mathematica* 23 (1957): 67–76.
Weisheipl, James A. "Curriculum of the Faculty of Arts at Oxford in the Early Fourteenth Century." *Mediaeval Studies* 26 (1964): 143–85.
———. "Developments in the Arts Curriculum at Oxford in the Early Fourteenth Century." *Mediaeval Studies* 28 (1966): 151–75.
Willard, Charity Cannon. *Christine de Pizan: Her Life and Works.* New York: Persea Books, 1984.
Wood, Chauncey. *Chaucer and the Country of the Stars.* Princeton: Princeton Univ. Press, 1970.
Zinner, Ernst. "Cl. Ptolemaeus und das Astrolab." *Isis* 41 (1950): 286–87.

Index of Terms

In the following entries, the pages on which each term is most fully explained are listed first, before the semicolon.

alidade 72 n. 47, 90 fig. 4; 24, 46, 47, 88
almanach 72 n. 51; 40, 52, 98, 99
almucantharat 36, 66 n. 5, 68 n. 25, 89 fig. 1; 16, 18, 24, 38, 42, 44, 46, 48, 50, 52, 54, 56, 84, 85, 94, 95
almuri 38, 71 n. 42, 89 fig. 2; 18, 24, 42, 44, 46 48, 54, 56, 84, 85, 90, 94, 95
angle 103 n. 13; 24, 32, 33, 85
annel 68 n. 17; 18, 34, 36, 40
armille 68 n. 17; 18, 24, 34, 90
ascendent 40–42, 73 n. 60, 103 n. 16; 20, 24, 32, 46, 48, 50, 54, 93, 95
azimuth 66 n. 5, 69 n. 31; 18, 36, 46, 48, 50, 84, 85, 88, 89
bort (="lymbe," q.v.) 36, 38, 42
bout de estoile 46
bout du dent 42
cenith (=zenith) 18, 20, 24, 36, 38, 48
cercles des heures 38, 75 n. 77, 90 fig. 4
cheville, cheval 38, 71 n. 43; 18, 24
crepuscle 73 n. 64; 22, 42
declinacion 76 n. 95; 15, 24, 48, 50
demy quarré 72 n. 45, 90 fig. 3; 19, 38

direct see "retrograde"
do(r)s 34, 90 fig. 3; 15, 19, 20, 38, 40, 46, 48, 49
ecliptique 38, 70 n. 39; 18, 24, 42, 52, 76 n. 92, 77 nn. 98 and 99
equinoctial 36, 89 fig. 1; 18, 24, 38, 50
estoile fixe 20, 34, 38, 40, 42, 44, 46, 48, 50, 52, 92
face, visage 34, 50
figure du ciel (=horoscope, horoscope chart) 78 n. 105, 100 fig.; 50, 52
firmament 10, 32, 34, 36, 38, 40, 92
hautesse, hauteur 22; 18, 20, 21, 32, 34, 38, 40, 42, 44, 46, 48, 50, 52, 54, 56, 58, 60, 62, 90, 92, 94
heure equale 40, 74 n. 68; 20, 33, 44, 54, 56, 70 n. 34, 92
heure inequale 40, 74 n. 68; 18, 20, 36, 42, 44, 46, 54, 56, 70 n. 34
(h)orizon 69 nn. 26 and 27; 18, 24, 36, 38, 42, 44, 46, 48, 52, 94
horologe 48, 54, 92
jour artificiel 74 n. 68; 44
latitude 76 n. 95, 77 n. 98; 15, 20, 24, 34, 52

113

lieu (du soleil, des planetes, de la lune) 72 n. 50; 20, 40, 42, 44, 46, 48, 50, 52, 94, 100
li(n)gne de midy 36; 18, 42, 46, 52, 54
li(n)gne de minuit 36; 18, 54
li(n)gne de vrai orient et occident 36; 18, 40, 50
li(n)gne de vrai orizon 36
longuete 89; 38
lymbe 68 n. 20; 18, 34, 48, 54
maison 77 n. 102, 102 n. 2; 21, 34, 52, 54, 92, 94, 96, 98, 99
mere 34, 68 n. 18; 18, 24, 38, 42
meridionel 70 n. 39; 20, 38, 52
nad(a)ir 42; 24, 44
nuit artificiele 74 n. 68; 44
oblique ascension 48
orizon oblique 18, 36
planete 18, 20, 21, 22, 34, 40, 44, 48, 50, 52, 54, 56, 92, 98, 100
practique 10–11, 32, 52, 56, 62, 65 n. 1, 92

proffis 32, 34, 56, 62
quadrant 46, 75 n. 78; 58, 62
reigle, riulle 38, 72 n. 47; 19, 40, 46, 48, 50, 56, 58, 60, 62, 90
retrogradacione 20
retrograde 52; 20
reys, rethe 70 n. 37, 89 fig. 2; 18, 20, 24, 38, 42, 44, 48, 50, 54, 56, 88
riulle see "reigle"
r(e)ont 22, 32, 34, 36, 50, 52
septentrionel 70 n. 39; 20, 38, 48, 52
signe 38; 19, 20, 32, 34, 40, 42, 46, 48, 50, 52, 54, 92, 94, 96, 98
stacionere see "retrograde"
table (plate, disc) 18, 19, 24, 34, 36, 38, 40, 89, 94
tablet (vane) 19, 38, 40
visage see "face"
zodiaque 76 n. 97; 20, 38, 42, 46, 50, 52, 54

Pèlerin de Prusse on the Astrolabe presents the text and translation, with introduction and commentary, of the *Practique de astralabe*, until now an unedited and little studied medieval French treatise on this ancient astronomical and astrological instrument.

Written by Pèlerin de Prusse in 1362 at the behest of Charles V (while he was still dauphin), the *Practique* is an important example of early technical and scientific writing in the vernacular and illustrates the court's intellectual occupations and abilities. Since the treatise is heavily dependent on the same source as that on which Geoffrey Chaucer relied for his more famous work on the same subject, it invites comparisons with Chaucer's work.

Edgar Laird is Professor of English at Southwest Texas State University; **Robert Fischer** is Professor of French at Southwest Texas State University.

MRTS

MEDIEVAL & RENAISSANCE TEXTS & STUDIES
is the publishing program of the
Center for Medieval and Early Renaissance Studies
at the State University of New York at Binghamton.

MRTS emphasizes books that are needed —
texts, translations, and major research tools.

MRTS aims to publish the highest quality scholarship
in attractive and durable format at modest cost.